David Thomas Ansted

Geological gossip

Stray chapters on earth and ocean

David Thomas Ansted

Geological gossip
Stray chapters on earth and ocean

ISBN/EAN: 9783337275921

Printed in Europe, USA, Canada, Australia, Japan

Cover: Foto ©berggeist007 / pixelio.de

More available books at **www.hansebooks.com**

GEOLOGICAL GOSSIP:

OR,

STRAY CHAPTERS

ON EARTH AND OCEAN.

BY

PROFESSOR D. T. ANSTED, M.A., F.R.S.

HONORARY FELLOW OF KING'S COLLEGE, LONDON;
LATE FELLOW OF JESUS COLLEGE, CAMBRIDGE;
LECTURER ON GEOLOGY AT THE E.E.I. MILITARY COLLEGE
AT ADDISCOMBE,
ETC. ETC.

LONDON:

ROUTLEDGE, WARNE, AND ROUTLEDGE,

FARRINGDON STREET.

NEW YORK: 56, WALKER STREET.

1860.

[The right of Translation is reserved.]

CONTENTS.

LIST OF ILLUSTRATIONS.

ADDENDA ET CORRIGENDA.

Pages 78, 79. In the brief outline of African discovery in these pages, a reference to Livingstone's first visit to Lake Ngami from the Cape in 1840 has been omitted, and the direction of Caillié's route is wrongly stated. Caillié reached Timbuctu from Senegal, and thence crossed the desert to Tangier.

p. 100, line 20, *dele* the word *smallest*, and add the following note :— "The Fitzroy, Glenelg, Prince Regent's River, and others, have been entered by boats to some distance, but do not appear to open out the interior of the country."

p. 128, line 9 from bottom, *dele* B.C.

GEOLOGICAL GOSSIP.

I.

WATER AND ITS CIRCULATION THROUGH AIR AND EARTH.

What water is—Where it is—What it contains—What is done with it by the air—What becomes of that part of it that falls on the earth—What becomes of it when it penetrates within the earth—How it gets out of the earth.

THERE is much to be learned even from the very simplest things in nature, and when we begin to inquire about such of them as are seen and made use of every day, it is astonishing how little is really known, even by well-informed persons, and how large is the field of observation. Without entering on any difficult problems, unsolved as yet by the chemist and the geologist, let us consider what is known about water and its mode of circulating through the air and earth so as to be available for the ordinary purposes of life. Perhaps in this way we may find some not unpleasant reading combined with some useful information.

1. To begin (without discussing the history of the great discovery of its compound nature), let us

B

inquire what water is.   It is composed of equal
quantities, or, as chemists express it, equal *volumes*, of
two important gases, oxygen and hydrogen—these
being probably the two most abundant and important substances in nature, as regards ourselves
and our earth.   These two gases, which cannot at
present be brought into combination without an
electric spark—which, when separate, have never
yet been obtained in a liquid, far less in a solid form
—which have never been seen, or felt, or tasted,
but which are everywhere about us, and form a part
of everything we see, feel, and taste—are, by the
means referred to, brought into a marvellous combination.   They become converted into vapour,
many gallons of them when thus combined forming
one small drop of fluid water.   Under ordinary temperature, and under the pressure of the atmosphere,
water, as we know, remains fluid on the earth's
surface, but when heat is applied or the pressure is
removed, it becomes vapour or steam, invisible* and
intangible, like the gases it came from, but still
perfect water. It can in that state be reduced back
to its elements by passing it over iron at very high
temperature, the iron then attracting the oxygen
from it; but the passage from, as well as to, its condition as water involves marked electrical changes.

Water may very easily be brought into the solid
form by removing heat.   In this state it is elastic,

----

* High-pressure steam is well known to engineers to be
invisible.

transparent, and hard. In cooling down from a temperature of 40° and entering the solid state, water slightly increases in bulk. In being heated from 40° towards the boiling point, after which it passes from the fluid into the gaseous form, it also increases slowly in volume. Once in the state of vapour, it increases very rapidly and is highly elastic.

Water is probably an universal solvent, taking up silica (or flint), lime and clay, iron and lead, besides dissolving more or less of all earths and metals, acids and salts. Wonderfully small, no doubt, is the quantity absorbed of many of these substances, and perhaps in some cases the dissolving at all only takes place when the temperature is high. Water contains, however, in its ordinary state in the sea or earth, sufficient solid ingredients to play a marked part in the changes that take place in minerals, often apparently turning one into another. Most substances are dissolved much more readily, and in larger quantity, by hot water than cold; but carbonic acid gas is contained in greater abundance the colder the water. This and other gases can also be forced into water by pressure.

Water is absolutely essential to life, both vegetable and animal, some even of the most minute animalcules being only active in water, though they retain dormant life for an indefinite period without it.

Water is a most powerful agent in wearing away and rearranging the materials of which the different

strata of the earth's crust are made up. In the
solid form it carries millions of tons of stones and
earth from one part of the earth to another. In
the vapour form it is STEAM—a word which expresses all that man has hitherto done that is
greatest, most useful, most powerful, and most
valuable in reference to brute matter. In the
liquid form water bears upon its bosom our ships,
and carries ourselves and our industry to the utmost
parts of the earth.

Such is water—the commonest and the most useful
material, the simplest and the most powerful agent,
without which there could be no living being, in
the sense in which we understand life, and which,
circulating about and through the earth, renders it
available for living beings, and gives it, as it were, a
vitality of its own. That there is no appearance
of water in the moon is proof sufficient that no
organization conceivable by us is possible there.

2. Where then is it that water is to be found?

Go where we will upon our earth, it is everywhere present. The great ocean — a body of
water occupying seven-tenths of the surface of the
globe—covers all its deeper irregularities to depths
varying from a mere film to thirty or forty thousand feet. The whole mass of the water, including
the Atlantic and Pacific, and the smaller oceans, is
perhaps equivalent to a complete coating of the
earth's surface, if it were perfectly smooth, having
a thickness of nearly a mile. Spread over wide

tracts, only occasionally interrupted by groups of
islands, it marks the outline of the land by an
irregular line, here forming an inlet, there an
inland sea, separating important islands from the
mainland, and almost dividing large continents.
At one time calm and peaceful, at another time
lashed into mighty waves and angry foam, it ever
undulates with the great tidal movement, swelling
outwards towards the moon as the earth revolves,
and it is gently swept or suddenly disturbed by the
winds that perpetually pass over its surface.

On the land there is also much water, most of the
great depressions on its surface, though not all,
being thus covered, and forming a number of large
and innumerable smaller sheets.

In addition to these are the rivers, pouring ceaselessly their contribution to the ocean, fed silently
by springs, and brooks, and streams, and at length
becoming large bodies of water, always pressing
onwards in one direction, all traversing the land,
most of them finding their way to the sea. Down
from the mountain sides come the melting snows,
rushing madly towards the lower ground, foaming
and leaping, instinct, as it were, with life. Soon
they enter the valleys, and become water-courses
of another kind, larger, more measured, and more
sedate. From point to point these receive con-
tributions from smaller sources, obtained from
lower mountains and hills. At length the stream
reaches the plains, and then toils on at a slower

and slower pace until it enters the sea, loaded, and almost choked, with the wealth it has accumulated, and depositing a tongue of fertile land stretching out beyond the original limit of the coast.

Within the earth, also, there is present in abundance this fertilizing and life-giving fluid. In the soil, whence springs a rich carpet of verdure or lofty forest trees—in the rock beneath the soil, supplying it with moisture after long seasons of drought— beneath the arid stone and thirsty sands of the desert—in all these, and a thousand other places, a little labour will show that water is present, and may be obtained by proper means. And even more than this, if we take the solid granite, the limestone, the sandstone, the stiff, hard clay, and examine them chemically, we shall find that, without water, even their solidification would have been impossible, and that each one contains its due proportion, and many of them a very marked percentage.

In the air, also, there is water in large measure, and the state of the air, except with regard to temperature, has little or nothing to do with the quantity. In the coldest and driest air, where it might seem least likely; after crossing the burning deserts of Africa or Australia, where it might seem impossible; and while sweeping along in the fierce simoom, carrying death with it; in all these cases there is water in abundance, easily detected by proper instruments. But in the air that comes to us fresh from the ocean, or that has passed over any large tract of cultivated or forest land, the proportion is not only large, but is often sensible to the senses. Let there be once a small change of temperature induced—let a glass of spring water be brought cool into such air on a warm day, and we see the drops mantling on the outside of the glass, the water detected and deposited by this simple contrivance.

Clouds also are water, and clouds and mists are ever forming and travelling through the air. Many of those who read these pages may have seen in a mountain country the sudden formation of a cloud by a rapid shift in the direction of an air current, or have seen a breath of air advancing towards a cold mountain side suddenly converted into visible vapour by the condensation, in the form of cloud, of part of the water it contained. Air holds water in proportion to its temperature, and when charged with vapour, whatever the temperature may be, if the air is suddenly cooled, it at once gets rid of some of its load.

3. Let us in the next place inquire what are the contents of this fluid?

To a certain extent it may be described as an universal solvent, whose real contents no one can tell — for we know little of the minutiæ of nature's chemistry—but it is easy to detect some of the solids it holds in solution or suspension, under ordinary circumstances, and on a large scale.

In ten thousand parts of sea water there are of
common salt 270 parts, of Epsom salts, 56, of
Glauber's salt, 47, of carbonate of lime, 13, of silica or
flint, 3, and of sundry matters 3 parts—in all 392
parts, besides gases, of which atmospheric air is the
most abundant.   All these can be detected.   The
iodine, iron, and other substances known to be present cannot be thus calculated.   These quantities
are not to be despised, for we find that, estimating
the average depth of the ocean at 5000 feet, the
total quantity of common salt would amount to more
than 30,000 millions of millions of tons, while that
of silica, small as the percentage seems, would be
500 millions of millions of tons.   But this is not all.
The fresh water also contains inorganic salts to the
extent of from two to three parts in 10,000, besides
carrying a special load to the ocean, or depositing
it in its course, and in some cases that load is of
real importance.   The Ganges alone is thought to
carry 7000 millions of tons of mud every year to
the ocean, and the Nile has long been accumulating
mud at its mouth, which, in the course of ages, has
formed that extensive delta to which Egypt owes
its existence, the earliest seat of human civilization, and a tract of land whose fertility is nowhere surpassed.   In other places, as at the mouth
of the Elbe, the mud thus accumulated consists
not so much of the material brought down by the
river, as of the remains of countless myriads of
organic beings killed where the contact of salt and

fresh water takes place. Thus the mud itself, part of which is known in some rivers to be drifted over several hundred miles on the surface of the ocean, and which is probably carried much further by the under-currents, is a record of shore life, and mixes with the almost similar heaps of the shells and cases of foraminifers which have recently been found to pave the vast depths of the wide Atlantic for the eighteen hundred miles that extend between the shores of America and those of Ireland.

In the earth we find water springing out in many places, sometimes at a very high temperature, abounding in various minerals, often in a state admirably adapted to restore health in obstinate disease. Mineral springs contain many salts, iron, sulphur, silica, and other substances, in a state of solution, and with them nitrogen, sulphuretted hydrogen, and carbonic acid gases.

Within the earth again, in mines, we find water, often warm, containing copper and other metals; and these are good indications that many of the most marked and peculiar phenomena of mineral veins are due to the passage of currents of heated water loaded with minerals in solution. Look at the caverns in limestone, gorgeous in their picturesque resemblance to gothic temples and varied imitative forms; all these are produced by the particles of carbonate of lime left behind when water has trickled through crevices in the rock above, dissolving a part of it, and has afterwards evaporated

while suspended from the roof, or resting in a
puddle on the floor. In a precisely similar way
are produced deposits of flint and chalcedony, in
romantic shapes, in the hot springs of Iceland and
elsewhere; but in these cases silica takes the place
of the carbonate of lime.

Water, then, contains large and important quantities of solid matter, which it either conveys
mechanically to a distance, or having dissolved
them at one time, gives them back again at
another. It is an agent capable of doing much in
the modification, and even reconstruction, of the
earth, because it acts universally, and is possessed
of great and valuable chemical peculiarities tending
in this direction.

4. What is done with water by the air?

The atmosphere floating evenly over the general
surface of land and water, as the ocean floats over
part of the earth, is subject, like the water, to the
special attraction of the sun and moon, and exhibits
tides, greatly modified, no doubt, by the heat of day
and the cold of night, by summer and winter, and
by electric and magnetic storms, but still very
sensible and producing marked results. The result
of all these forces acting on the air is the creation
of incessant currents of wind—the air as it passes
over warm seas taking up as much water as it can
carry, conveying this portion by degrees to the air
above until it becomes saturated, and thus removing in the state of vapour daily, and hourly, if

circumstances are favourable, a quantity of water, which is afterwards distributed over the surface of the land, and serves to produce that circulation of fluid which is, as it were, the blood of the earth, resembling as it does the circulation of the vital fluid through the animal frame.

The quantity of water thus lifted is exceedingly large—far larger than could at first sight be supposed. In warm latitudes, in the Atlantic—and no doubt also in the Pacific—the lower portion of the atmosphere constantly absorbs and retains in an invisible form an enormous quantity of vapour, and these lower strata gradually arising to the higher and cooler parts, there condense into cloud. In that condition they are drifted steadily onwards, at an elevation of 15 or 20,000 feet, by a steady South-west wind, which blows during the whole of the summer, if not all the year round, from the whole inter-tropical region of the earth. To meet the vacuum caused by the incessant lifting up of this great body of warm air loaded with vapour, there is a steady North-easterly current of dry cool air, which sets in from the Continents of Europe and Africa, at an elevation of between 2000 and 3000 feet. Below this again are the variable and shifting winds, of whose proverbial inconstancy we are all well aware.

When the warm winds reach the land, they are subject to many influences, of which a change in the electrical elements is perhaps among the most important, and the clouds then break up, and sooner or later deposit their load of water.

In some places the quantity of water dropped on the earth in this way is but small—equivalent, during the whole year, to a uniform vertical sheet of from sixteen to twenty-four inches thick; but in other places, especially in warm latitudes and the North-western coasts of Europe, where the South-west wind first strikes the land, as much is received in one year as in other places in five or six years. It may seem that this does not amount, after all, to a great deal; but when put into other words, and we learn that each acre of ground receives for every inch of rain that falls on it upwards of 22,500 gallons of water, which all has to be accounted for in some way or other, a more distinct idea of the importance of the supply will be obtained. It is thought that the average amount of rain-fall over all the land of the earth is not far from thirty-six inches, or is equivalent to a stratum of water a yard thick covering the whole surface. It is also calculated that an acre of cloud one hundred yards thick, at a temperature of 70°, contains 2500 gallons of water, half of which would be discharged if the temperature were suddenly reduced to 40°, so that each acre of such cloud raining upon the earth, under the change supposed, would deposit upwards of a twentieth of an inch of water—and if such a cloud were moving through the air at the rate of ten miles per hour, it might rain down an inch deep of water in a few minutes. Such a shower, however, would be of a kind only seen within the tropics. The air is never really deprived of all its moisture during the heaviest rain, always retaining the quantity proper to its temperature at the time; and when the temperature is afterwards raised, it immediately again absorbs. The part not deposited on the earth drifts away, and is subject to fresh changes.

5. What then becomes of that part of the water that falls to the earth?

This question is a very important one, and involves a multiform answer. In the first place, it appears that almost immediately after parting with any quantity of its moisture the temperature of the air rises, and a part of the water is re-absorbed. Whatever is not thus re-absorbed necessarily rests on the ground, which of course varies greatly in different places, here being tough clay, there open porous soil, in another place hard rock, and occasionally loose sand. The water, according to circumstances, either remains on the top in pools, soaks into the top, leaving it spongy, or sinks altogether below the surface. When afterwards a drier and warmer wind blows over the surface, this wind will absorb moisture very easily and rapidly from the pools, and easily enough from the spongy surface, but not so rapidly, and sometimes scarcely at all from other parts where the water has penetrated crevices and passed down into the rock, or has sunk far below into loose open beds, or into others more compact and removed from surface influences.

But whenever the surface is damp, a part will be re-evaporated, and out of the whole quantity of rain that falls, nearly two-thirds are estimated to be thus taken back again into the air. The proportion, however, varies in every place, at all seasons, and at all hours of the day.

It also follows from the irregular form of the land, as well as from the different kinds of rock on which the rain falls, that a very large part runs off from the immediate surface to form those numerous streams and rivers that drain the surface and render it habitable and wholesome. The principal rivers running into the sea are calculated in temperate climates to carry back about one-sixth part of the whole rain-fall; and perhaps in the tropics, where the rivers are some of them very large and rapid, the proportion may be even greater.

We must not overlook another great cause of consumption of water, both of that which soaks into the soil and that which collects in pools or runs along in rivers—namely, the supply required for the animal and vegetable world. When it is remembered that a very large proportion of the weight of every living being, animal or vegetable, consists of water, and that for life to continue at all, an incessant supply of fresh fluid is required, the necessity of water in the air and on the earth will be easily understood. The absorption of water by vegetation, and its conversion into leaves and wood in the growth of plants during the spring of the year, cannot but be looked on as a most essential agent in the removal of a large part of the rain by locking it up in organic substances. Of each ton weight of growing grass, very little more than two hundred weight can be obtained as dry hay; and though this proportion of water in the weight of growing plants is by no means to be regarded as an average, it is safe to calculate that one-half, at least, of all organic matter consists of water, while in a living state.

Although a large quantity of the rain that falls is used at once by animals and vegetables, and a much larger quantity is re-evaporated into the air or drained off by rivers into the ocean, there still remains a large residuum, which must next occupy our attention. This portion disappears for a time, being apparently lost and buried in the earth. It is not lost, however, for though away from observation, it is still proceeding on a useful course— fertilizing the earth above—stored in deep natural reservoirs for a future day—carried through and modifying rocks far removed from the surface, and after running its course, coming back once more and reaching the sea to perform in time another circuit.

A large part of the rain water that thus sinks into and travels through the earth, issues forth again either from natural springs or by artificial openings; sometimes welling out at the bottom of ponds or lakes of water, or on the seashore, at a great distance from its source of supply—sometimes resting for a while in deep underground pools, which require that wells should be sunk in the earth to let it pass, and occasionally bursting out with violence at a high temperature, or loaded with mineral salts. The part of the water that thus follows a subterranean course is not only the most obscure and difficult to trace, but is the longest in circulating, and changes much more than any other part of the water from the simple and pure state in which it exists when first distilled from the sea.

6. We have next to consider what becomes of the water when it enters the earth?

All soils are more or less porous—few absorbing less than five per cent., and some as much as forty per cent. of water. We may regard the surface of the earth, wherever there is vegetation, as a kind of sponge, sucking in a large quantity of water and conveying it to the subsoil, and to the surface of of the rock below. Now, if any exposed face of rock is observed carefully, no matter how hard the rock may be, or how large and solid and un-flawed the slabs that may be quarried from it, we shall find the weathered or exposed face split and broken into ten thousand fragments close to the surface; and besides these cracks, very numerous systematic crevices will be recognisable, breaking up the mass into cubes, or other solid figures of definite form.

When the rock is stratified, or arranged in layers, some of which allow water to percolate them more readily than others, the water will accumulate in quantities on the top of those that are least permeable, and run along them till it meets some of these crevices. Where, on the other hand, the rock is hard but also cavernous, the water will collect in the cavities, and where the rock is not only hard but soluble, the water will gradually wear away hollow spaces and passages. It will, in a word, circulate among all rocks, whether stratified or unstratified, whether hard or soft—will penetrate into almost every part even of the least porous rocks—will exist sometimes in open cavities, and sometimes, under great pressure, in crevices to which the day never penetrates, and of whose position no one can guess who is not endowed with the peculiar second sight of the *dowser*.* In sandstone rocks and chalk, actual experiment has shown that while the water is abundant in many parts of them, and certainly travels through their substances, it yet advances very slowly through the mass of the rock, though rapidly enough where it runs along open crevices. Even in loose sands and gravel, however, the passage of the water is by no means rapid, while in clay it can scarcely be said to pass at all.

---

* Persons so called are to be found in Cornwall and elsewhere, who profess to be endowed with a kind of second sight, or sixth sense, enabling them to be aware of the presence of springs of water, and sometimes of metals, beneath the earth, by the twisting of a hazel stick, held in their hands, while walking carefully over the surface.

Almost the whole of the rain-water that enters the earth, whatever the quantity may be, penetrates the rocks below the surface. It does so to a variable, though perhaps nowhere a very great depth, being conducted along through underground channels and passages, often narrow and curiously contorted, trickling down the splits and crevices of the harder rocks, or gliding along on the surface of the tougher ones. Where interrupted by a belt of impermeable ground it will accumulate—where conveyed to the open air it will run off, obeying in all cases the law of gravitation. Occasionally, where the depth to which it passes is considerable, this is shown by the equable and often high temperature at which it rises ; and this is not unfrequently the case where there is nothing in the contents of the water to indicate that it has undergone an essential change.

7. How is it, lastly, that water gets out of the earth ?

This question is a practical one, and requires consideration accordingly.

When water reaches to the bottom of the surface-beds that are permeable, and collects in hollows there, owing to the uneven surface of the water-bearing rock below, if these surface-beds, which we may regard as sand or gravel, are themselves uneven, or rest on a sloping foundation, we shall occasionally have springs obtainable by digging a well into the hollows at the bottom of the gravel itself, or at the point where the gravel capping terminates.

If there is no natural opening, and the circumstances are such that the water is forced to accumulate till it rises to some underground outlet, then our artificial well, although in dry seasons occasionally emptied by pumping, will be constantly refilled every shower, at a faster or slower rate according to the nature of the gravel and sand, and the facility it offers for the transmission of water. Wherever the rock immediately beneath the soil consists of any thickness of gravel, or rolled blocks of stone, or of fragments of rock, allowing water to pass freely between and amongst them, there we may expect to obtain water near the surface, derived from this source; but we must also expect that in a dry summer, or if any quantity of water is removed at all approaching to the quantity supplied by the rain-fall of the district, the supply will fail, sooner or later. Such supplies may exist at any level, and are as often at the tops of hills as in valleys. They are very common, and only require that the water level should not be too deep to be accessible. Wells sunk in such material are not generally very costly, and may be repeated in any part of the deposit, and it will often happen that the quantity, and even quality, of the water will vary a good deal at different points.

Where water does not well out of the earth from such surface-beds as gravel, we may still often find it where a clayey rock is intersected on the side of a hill, provided there is a permeable rock above it. For this, however, certain geological conditions are necessary. Either the water-bearing bed must be quite filled and choked with water, or the bed must incline towards the spring and receive a supply of water from some higher point at a distance. Springs from a hill side are generally of this kind; and if there is any natural impediment to the passage of water down a bed shut in above or below by some tight rock, the water will accumulate against the impediment and tend to remove it. If the impediment is merely soft mud, this is gradually removed, and the water forces its way upwards to find its level. Such is the case when beds are faulted or broken, and the fracture filled up with clay, and in this way natural springs occur at faults and on hill sides. If, however, there is any thing to prevent the spring from rising naturally, it is often easy to make it flow by artificial means; and thus by a well sunk a few yards into the earth a source of water may be reached which is not so immediately dependent on the season as are the land springs, although it also is fed entirely by the rains, and is often dependent on the average of several successive seasons.

Some limestones and some sandstone rocks, are hard, but exceedingly cavernous, or full of holes and crevices. Some, as chalk and soft sandstones, are very porous, and suck in a great deal of water, which they retain like sponges, and will subsequently give out. In these the water sinks to the bottom of the rock very slowly, and there it collects in a body, and is capable of being pumped, especially if a lodgment or cavity be artificially made.

However wet such latter rocks may be, the quantity of water that can be pumped from them is limited, and depends on the nature of the rock. Thus, from one to two millions of gallons daily can be pumped from a single well in porous sandstone, and somewhat more from chalk; but hardly more, however large the extent of rock or the supply of water. The mode of obtaining water by pumping from such rocks continually has sometimes been adopted for the supply of large towns; the wells sunk for this purpose, which are often very deep, being called wells of exhaustion, and draining the rock within a circle of one or two miles round, or somewhat more. Over the whole of this area the surface and shallow wells will more or less suffer.

Occasionally much water is obtainable by penetrating through certain rocks which do not allow water to pass, and thus reaching others at a considerable depth, which are permeable, and which are fed from a distance. Springs may in such cases rise not only as high as the surface, but even much above it, if they are fed from a much higher level than that at which they discharge. They are called *Artesian Springs*.

When water penetrates limestone rocks, and occupies the cavities, these can sometimes only be discharged by tortuous passages at various levels, and occasionally it happens that the level of part of the delivery pipe is above that of the reservoir at ordinary heights. In this case the delivery pipe acts as a syphon tube, and if the outer leg is longer than the one nearest the reservoir, the water will not be discharged at all until its level rises above the bend of the syphon, but then the accumulated store will be all emptied. In this case an interval will elapse when there is no spring, and the water will be discharged intermittingly—such intermittent springs are not uncommon.

Lastly, water is discharged at natural openings, either under strong steam pressure, as near volcanoes, or by hydrostatic pressure, and often at a very high temperature, as happens in mineral springs, of which there are many discharging large bodies of warm water. These waters often contain mineral salts, and even metals or silica, together with a large quantity of gas derived from the various rocks and materials through which they pass.

Such, then, is a brief outline of the natural history of water—the simplest of combinations, and the compound most resembling a simple element—the most universal solvent at all temperatures—the most widely distributed substance in nature—the most powerful agent—the most perfect representative of perpetual motion, penetrating everything, passing everywhere, always present, in sight or out of sight, and everywhere producing a marked effect.

The sea, a vast receptable of moving water, holds in solution a large quantity of various minerals, and is a seat of life of whose range and importance it is difficult to have too large an idea: while the rivers carrying down to the sea treasures of mud and solid matter held in suspension, deposit a great part of their load where they come in contact with salt water.

From the great ocean rise into the air those vapours of water which, as they rise, become clouds, and are drifted along, perhaps for thousands of miles, till they reach high land. Arrived there, or interfered with by any cause, they deposit their watery load as rain which falls in drops on the earth.

Of the rain thus thrown down, part falls on hill or mountain sides, and collecting in gullies, soon accumulates and forms mountain torrents. These as they proceed along an accustomed road back to the ocean increase in volume by the addition of numerous tributaries, and diminish in speed, until they become fertilizing rivers, either meandering through vast plains, or rolling on in their majesty between massive walls of solid rock.

Another part enters into and forms a constituent of animal and vegetable life which cannot continue to exist without the refreshing supply.

Still another part has very different duties to perform. It sinks within the earth and is received among the rocks and strata beneath, as into a reservoir, and then, at a high temperature, and loaded with numerous mineral salts and simple minerals, it circulates through rocks, destroying some, altering some, and forming many, until at length, its numerous tasks performed, it comes forth once more to the welcome light of the sun, trickling slowly in small streamlets from a hill side, welling freely from a natural fountain into a river or lake, or bursting forth with violence under the strong pressure of steam, as in the boiling springs of Iceland.

Such being the course taken by water in its transit through the air and the earth, let us consider for a moment what would be the contrast if on the one hand the temperature of our globe were so far diminished as to lock up all the fluid water into solid ice, or, on the other hand, if owing to the absence of an atmosphere, or the increase of temperature, the water existed entirely as vapour.

Without water, that wonderful arrangement of strata upon which depends all that is beautiful and much that is essential for life in the disposition of the materials of the earth's crust could not have existed, for everything in that arrangement involves the constant and direct action of water, chemically as well as mechanically.

The face of creation would have been in a most important sense void, and might have been fairly described as chaos.

We may safely say that in a planet so constituted there could hardly be a single object presented to our vision, either familiar to our experience, or in any ordinary way comprehensible. There could be no such life as we live, no such incessant change as gives life now, even to inanimate nature, no such change as seems to us a part of the natural order of creation. The earth would present one dreary blank, in which the intensest glare of sunshine would alternate only with the intense blackness of perfect night. Such would be a world without fluid water and without an atmosphere of elastic gases.

But what if, in consequence of an increase of heat, the water of the ocean and of all rivers and springs should be evaporated and form a dense veil of visible steam or fog, or an atmosphere of invisible steam over the whole earth. In many respects the results would resemble those in the case just mentioned. There would still be no aqueous deposits, no circulation of water, no life like our life, no beings like those we are familiar with. Death would reign everywhere, and silence and stillness would take the place of the universal movement that now characterizes our earth.

In any case it is clear that our world has been adapted to its inhabitants, and the inhabitants to their world. This mutual dependence meets us everywhere, and evidently forms part of the great plan of creation.

II.

RIVERS AND WATER-FLOODS.

Mode of action of Rivers in eating away the land—Rivers of the South of Spain—Rivers of India—Ordinary bed and rainy season bed—Occasional torrents—Evidence of erosion—Recent period during which the erosion has taken place—Effects produced on climate.

AMONG the numerous agencies tending to remodel the earth's surface, the running water of rivers and smaller streams is one whose value has always been recognised, but of which the great importance in tropical countries, where the rain-fall is large, and the material and form of the surface are favourable for its action, has often escaped attention in geological calculation. The reader is probably acquainted with the illustrations of the magnitude of this force in Lyell's *Principles of Geology,* and other popular works on the same subject; but there has recently been published an account of the condition of Northern India in this respect, containing some details less known and far more striking than any hitherto recounted. Much also may be learnt from the similar operations on a smaller scale in the

South of Spain and in Northern Africa, also but little known.

Rivers act in various ways upon the land over which they pass. Their power of conveying mud, sand, or stones, and the distance to which they can carry material once moved, depends very greatly on local circumstances, but generally torrents only move material a comparatively short distance at one time; and in their course towards the sea are much more occupied in eating out new channels, obliterating old ones, destroying ancient banks and beds, and constructing new ones, than in actually delivering any particular part of their load at the sea-coast. There is, however, wonderfully little resemblance in the action of running water as observed in our own country and as noticeable in others differently constituted in the South of Europe. Still less is there any resemblance when the phenomena are investigated near some great mountain chain at whose foot or on whose flanks is a vast range of country, composed of material easily moveable and subject to torrents of rain, and to the subsequent inundations inevitable from the form of the land.

In a work published in 1854,* the author of the present work described some of the peculiarities of the drainage of the South of Spain, and especially those remarkable river-courses, often of considerable

---

* Scenery, Science, and Art, p. 154.

width, and consisting to a great depth of the broken fragments of the rocks adjacent. These river-beds are generally quite dry, or traversed only by a stream a foot or two in depth, and even when most swollen the stream does not occupy a large part of the bed, but rushes along in a rapid torrent destroying all before it, and often converting thousands of acres of the most fertile land into a desert.

In these cases the shifting of the bed of the stream—an operation happening with more or less violence every year—is the chief apparent change; but when these streams are examined carefully, and traced in their course through rocks capable of rapid destruction, they are easily proved to undermine their banks. They thus cause the fall of long lines of shingle and sand, by eating deeply into the sands and stones below—the accumulations of former years. The great sheet of gravel and loose boulders always left in a wide belt spread out on the plain where such water-courses approach the sea is thus discovered to be the inevitable result of the periodical action, when swollen by rain, of the contemptible streamlet that trickles along during the warm season.

But if in Spain the erosion of small streams is thus marked, it may be imagined that in India the large rivers that rush down from the Himalaya Mountains will be on a grand scale. This has long been known, but no accounts received are so striking as those by the travellers H. and R.

Schlagintweit, read at the Dublin meeting of the British Association, in 1857.*

In the northern part of India the number of feeders to the gigantic rivers, the Ganges, the Indus, and the Sutlej, is exceedingly large, and each one does its own work separately. Afterwards two or more of such feeders combine, and this is repeated till at length the vast torrent of the river is formed in the valley, and makes its way to the sea with gradually decreasing velocity, leaving behind at various points the material it has carried along during its earlier and more violent progress.

In central India the streams running towards the north have two distinct beds, one for the average height of the water, which is occupied permanently, while the other, during the three months of the rainy season, forms an extraordinary but not irregular channel to carry off the water. Thus where the narrower bed of the Ganges and the Jumna is for a great distance about a mile in width we find on each side of this narrower bed a broad flat belt more than a mile wide. This tract of land is often cultivated and bears rich crops after the rains are off, but during each rainy season it is entirely covered with water, the river then occupying the wider channel of three or four miles, instead of the narrower space only one mile across. In proportion as the rivers are smaller, the

---

* Report of British Association Meeting at Dublin, 1857. *Transaction of Sections*, p. 90.

rainy season bed becomes larger, and in ordinary years this extra bed is sufficient to prevent a mischievous inundation. But when the floods are more than usually heavy, and an inundation occurs, the extra bed ceases to be sufficient, and the Indus has been known to occupy a width of seven miles in a place where the usual river bed was not more than two. Such events are fortunately rare, for the misery and destruction they cause is incalculable.

Considering now the result of the ordinary seasonal increase in the water of the rivers, it appears that in India and Tibet the erosion of the upper valleys, and even the production of large valleys by this power of water, is carried on on the grandest scale. It would seem that in former times waterfalls and lakes have existed, but these are only possible where the erosion has not yet reached a certain maximum. The lateral valleys have now become so nearly equal in level in their lower parts with the principal valleys, that the waterfalls are at an end, while the lakes have also been emptied by the constantly progressive erosion, and their beds may now be seen quite dry.

In the plains the rivers have eaten away the ground over which they run to a depth varying from eighty to one hundred and twenty feet, and generally present, as already explained, two distinct beds, one for the dry, and the other for the wet season.

In the mountains the evidence of water action is, of course, different from that deep, simple erosion noticed in the plains. Thus spoon-shaped hollows are eaten out in the walls of valleys, these walls consisting of a detritus partly different from, and partly identical with, that still in the river, and connected lines · of conglomerate beds with sand and freshwater shells are seen along the steep walls of the valleys. All these simply mark places where the river has once been, and enable the traveller to estimate the water action. Judging from them it is found that even in the small valleys the depth of the erosion has amounted to from 1200 to 1500 feet, frequently reaching 2000 feet, while in the upper course of the Ganges, the Sutlej, and the Indus the bed of the river has been at one time as much as 3000 feet above its present level. This is proved by the presence of a stratum of that vast thickness, partly of solid rock, partly of alluvium, which has been in some places removed after having been cut through by successive torrents.

It is important also to remember that such vast results have been obtained by the action of rivers during a recent geological period. All the great deposits of Indian tertiary geology, and even of the drift and other more modern periods, must certainly have taken place since, and there has been no change, as far as we know, in the animal inhabitants of the country. At the same time so great an alteration may well have affected the climate by modifying the distribution of moisture and altering the range of glaciers. It is indeed hardly possible to estimate fully what may have been the results in altering the climate during this period; but there can be no doubt that very considerable changes must have taken place in this respect, since lakes and other pools of water are dried up, and a river which once swept over the surface now rushes through a narrow gorge.

III.

THE SURFACE OF THE ATLANTIC.

Distribution of oceans—The Pacific and Indian Ocean—The Atlantic: its length and breadth—Its dependent seas—The Mediterranean—The Gulf of Mexico and other inland seas— The tides in the Atlantic—The Gulf Stream—Other stream currents—Theory of the Gulf Stream—The Sargasso Sea—Prevalent winds—Cyclones and storms—Dust-rain, and its origin— Icebergs.

THE waters that cover the earth are collected into two great basins, communicating with each other at the two extremities, but of very different form and extent. The Pacific is an oval shaped basin, having one smaller and subsidiary ocean (the Indian Ocean) connected with it. The main body of the Pacific, between the two Americas on the east and the coasts of China and Australia on the west, occupies a space of some ninety millions of square miles, and the Indian Ocean extending westwards to the shores of Africa, and almost detached by the long string of islands of the Indian Archipelago, occupies another twenty-three millions of square miles. The Atlantic is an elongated trough-like basin of somewhat zigzag form, enclosed by Europe

D

and Africa on the one side, and by the east coast of the two Americas on the other, and occupies in all only twenty-seven millions of square miles. Small and comparatively narrow as it is, the Atlantic, from its position in relation to civilized countries, has always been regarded as the more important ocean, and it is consequently much better known than its sister expanse.

The length of the Atlantic trough, or canal, is sometimes described as amounting to ten thousand miles; but since, for as much as three thousand miles of this length it is not fairly enclosed by land on each side, we will not estimate it at more than seven thousand. Its greatest width, between Florida and the coast of Morocco, is four thousand miles; but the distance between the land of the Eastern and Western Hemispheres is in some places less than 1000 miles, and near the equator is not more than 1800. There is, therefore, some reason for the name of canal applied to it.

The Atlantic is shut in towards the North and South by ice. In the North the ice reaches the land on each side during the whole of every winter, and, indeed, for the greater part of every year.

The results of the recent polar voyages, first to discover that Arctic passage which had been so long suspected, and then to seek for some indications of our countrymen who had fallen in their day of greatest triumph, martyrs to the cause of science: —these results are sufficient to show that a water communication exists between the two great oceans by way of the north pole of the earth; although the extreme cold of that region retains the water almost perpetually in a solid form. There is nothing, however, to prove that an open water communication may not always exist under the ice from the Atlantic to the Pacific.*

Confining our attention to the Atlantic, we find that it has many subsidiary seas. The Mediterranean, sacred to every classical and romantic feeling inherent in civilized man, on whose shores were the successive seats of government of the earth for thousands of years, whose waves wash the shores of Palestine and Egypt, of Greece and Italy, on which all the early discoveries and early hardships of navigation were experienced:—this great sea—the cherished "central ocean" of the ancients, is of all the most interesting. Occupying nearly a million of square miles, it only communicates with its parent ocean by a channel a few miles in width. It is therefore little affected by what takes place outside, and may be regarded as in most respects an independent basin.

---

* The discovery in Behring's Straits of whales that had been not long before unsuccessfully harpooned in Baffin's Bay, was a curious proof of the existence of an open passage before the discovery of that passage by navigators. Whalers are in the habit of marking their harpoons with a date and the name of their ship, and in one or two instances, a very short time only elapsed between the date of capture of a fish in the Pacific and the harpooning it in the Atlantic. The breadth of ice that can be travelled under by the whale cannot be very great, as these animals require to come frequently to the surface to blow.

The Gulf of Mexico and the Caribbean sea form together a basin double the size of the Mediterranean, and not without many points of deep interest. If from near the narrow Straits of Gibraltar there issued forth those few small ships that bore Christopher Colon on his great voyage of discovery, it was an island of the Gulf of Mexico that first received the faithful navigator and his despairing crew. Those inland seas, also, as we shall see presently, are the parts of the earth to which we must look for much that affects the climate of the coast of Europe, and to them is due that almost separation of Northern from Southern America which has influenced greatly the progress of events in the development of those two countries.

There are other deep inlets of the Atlantic of scarcely inferior importance. The Baltic and White Sea on the European side, and Hudson's and Baffin's Bay on the west, are partial elongations of the body of the Atlantic, producing results greatly affecting the course of human events. The gigantic masses of ice that alternately form on and break away from the shores of Baffin's Bay drift down into warmer latitudes, and act an important part in modifying the climate of Northern Europe.

Omitting for the present any special consideration of the subsidiary waters, let us confine our attention to the main body of the Atlantic. Its waters are well known to be exposed to great disturbances of various kinds. They are subject to the semi-diurnal tidal wave in its ebb and flow, to other waves similar in appearance, but really only heaving the water upwards and sinking it downwards, and to those streams and currents which affect limited portions of the water, and which resemble rivers passing quietly on through tide and storm, ceaselessly performing their course, and always effecting their purpose.

The tide wave in the Atlantic is a movement in mass of the whole body of the water, which advances from the South towards the North during a period of six hours, producing a total average rise of a very few feet, and then retires southwards again at the same rate and to the same amount. Simple as this statement may seem, only contemplate for a moment the grandeur of the result—all the water of this vast trough seven thousand miles long, and averaging two thousand miles wide, being at one moment still, at the next starts majestically into motion, and as if endowed with life presses onward, and by slow degrees lifts up its whole mass, till after an hour it is some four or five inches above its original level. Hour after hour this continues, till after about six hours have elapsed, a maximum in height is reached, and the whole water raised in open ocean some three feet. The advancing wave is then checked and stops, and soon becomes a receding wave at a similar rate. Twice in every twenty-four hours does this marvellous alternation of level occur, and as it affects the whole body of the water, its results near shore are greatly affected by the narrowness of the channel and its form, so that the elevation of three feet is in some places multiplied into seventy, and in others reduced so as not to be observable. Who has not sat by the sea side watching the ceaseless undulation of the water, and trying to discover how it is that while there seems a rapid succession of waves coming in towards the shore, the water is really gradually ebbing away from his feet, and drawing off further and further any piece of wood or other floating object he is watching? How many also have experienced, while deeply interested in some pursuit among the rocks or in the caves, that insidious and sometimes rapid rise of the whole body of the water, cutting off retreat, though there is hardly a ripple to be observed and nothing to mark the change that is taking place. The mysterious obedience to the command, "Thus far shalt thou go, and no farther," might seem to indicate the direct presence of some supreme power, had we not elsewhere ample proof that this is no interference with any of those great laws imposed on matter from the beginning, and that the perfection of the Lawgiver is best seen in the absence of any need for occasional interference.

The tide-wave, though one of the most interesting and difficult to follow of the Atlantic phenomena, is not, of course, confined to the Atlantic.

It seems to originate in the vast body of open water in the Pacific, and is then forced onwards into the modifications just described, owing to the form and limit of the channel it affects.

The Atlantic is also traversed by several rivers of salt water. These cross it unaffected by the tide, except in so far that they are lifted upward and sunk downward by this great wave, but their course is independent. Such a river on the grandest scale is the great Gulf Stream—a mighty flood, pouring forth its ceaseless volumes of warm water from the Gulf of Mexico towards the Arctic Seas. These waters being warm, and therefore lighter than those of the ocean into which they pour, and which form their banks and bottom, do not readily mix, and for a long distance the eye can readily discern the difference that exists between them, the colour of the warm waters of the stream being of a deeper and more indigo blue than that of the waters of the ocean, which are more usually green in the vicinity.

The Gulf Stream, where it issues from the Gulf of Mexico, is not more than about thirty miles wide, and is believed to be somewhat less than 400 fathoms deep. It proceeds northwards, expanding and shallowing as it goes. At first it moves at an average rate of about five knots per hour, and travelling northward and eastward, is turned across the Atlantic, just grazing the Banks of Newfoundland. At this point the difference of temperature between its waters and those of the ocean it traverses, is as much as from 20° to 30° on an ordinary winter's day. Much farther on in its course—midway across the Atlantic, and even as it approaches the land of the Old World —this water still retains a comparatively high temperature, and beyond all doubt warms the air immediately above it to an extent which greatly influences the climate of Europe on which that air blows. If anything were wanting to prove the vast influence of oceanic currents on the temperature of land, the comparison of the climate of Liverpool with that of the island of Newfoundland (which is indeed somewhat to the South), or of Norway with Spitzbergen, would be sufficient illustrations to refer to. "It is the influence of the Gulf Stream upon climate that makes Ireland the emerald island of the sea, and clothes the shores of England with evergreen robes; while in the same latitude, on the other side of the Atlantic, the shores of Labrador are fast bound in fetters of ice."[*]

But it is not alone the absence of the Gulf Stream that produces this result. Whilst a current of warm water crosses from the Gulf of Mexico to the shores of Northern Europe, another current of water comes stealthily along at a considerable and increasing depth, from the hardly-melted winter ice of the Arctic circle towards the equator. This polar current passes below, and crosses far out of

---

* Johnston's Physical Atlas. *Streams and Currents.*

sight the Gulf Stream; but before doing that it has had time to cool down the eastern shores of America, and render the contrast between them and the European land in the same latitude more striking.

These currents alter their position with the season. In winter the Gulf Stream passes more to the East, forced to take that direction by the increased volume of the cold water from the Arctic Seas. After a hot summer the case is reversed, and the stream approaches the land more nearly before it begins to cross.

A third current, though of smaller importance, is known to set northwards round the interior of the Bay of Biscay to the western shores of England; and a fourth, also commencing on the outskirts of the Gulf Stream, sets southwards towards the coast of Guinea. The waters of the South Atlantic appear to set continually across from the seas near the Cape of Good Hope to the east coast of South America, before entering the Caribbean Sea.

It is not easy to find a complete and satisfactory explanation for these remarkable phenomena. They are supposed to originate in prevalent winds, and these, no doubt, may act in forcing the water to a higher level on certain shores whence a current sets to restore the level thus lost. But this is not sufficient to explain the varied and marked appearances connected with the great streams. Lieutenant Maury, U.S.N., to whom we are indebted for an interesting work on the " Physical Geography of

the Sea," has expressed an opinion, supported by
many facts observed by himself and others, that
the rotation of the earth from west to east, acting
in some measure independently on the waters,
which do not hold together as the solid earth does,
must produce a current in the same direction. The
replacement of the water thus removed by cold
water from the Arctic circle can be readily admitted to be by a deeper current partly out of
sight, and away from immediate recognition ; and
knowing as we do that there is such a current
setting southwards, and that the temperature of
deep water throughout the Atlantic is very cold,
this theory is further supported.

Whatever may be the cause of the Gulf Stream,
there is no doubt as to the effect it produces, and
we know how completely the whole aspect of the
vegetable and animal world in the Northern Hemisphere is affected by it. Once it would seem the
stream flowed through what is now the valley of the
Mississippi towards the Arctic circle. If so, it is
not difficult to believe that ice and snow may then
have prevailed over Northern Europe, whose climate
must at that time have resembled that of the gloomy
and uncultivated lands on the coast of Greenland and
Labrador. It would be difficult to say how small a
change in the direction of this great distributor of
heat would modify and injure the climate of England.

On a large part of the Atlantic, nearly midway
across between Portugal and the west of North

America, is a curious expanse of sea, generally covered with a particular kind of sea-weed.

This would almost seem to be the result of a kind of eddy on a gigantic scale, into which certain marine vegetable productions in a living state have been drifted. At any rate, it is certain that this portion of the ocean is permanently and thickly covered with such growth. It is called the Sargasso Sea, from the name of the weed, and although some accounts given of the abundance of weed seem almost too wonderful to be credited, and it is not unlikely that in many cases the floating portion varies and breaks up for a time, there remains no doubt of this fact, that there is here collected in the very midst of a great ocean, far removed from land or shoal water, a vast heap of vegetable matter, and that no other similarly furnished tract of open water is known to exist.

Not less remarkable for its clouds, its winds, and its storms than for its tides and currents, the Atlantic offers in all these respects groups of phenomena worthy of the most careful study.

It is only in particular latitudes, and within limited distances from the earth's surface, that the winds may be quoted as typical of unsteadiness. Over a large part of the Atlantic, they blow with great uniformity in the same direction, but this direction is not the same at the respective heights of ten and twenty thousand feet above the earth. Generally it is found that near the Equator, and

near the line of the tropics of Cancer and Capri-
corn (seen marked on every globe), there are belts
of calm and of irregular winds.   Between the
equator and each tropic are regions in which the
air always blows steadily onwards from the east
—becoming north-east winds north of the line,
and south-east south of it.   This regularity does
not, indeed, always reach down so far as the surface, but prevails at a moderate elevation throughout the Atlantic, with extraordinary steadiness; and
beyond the belts of calm in the two temperate
zones there are other wide belts over which winds
blow much more constantly in some one direction
than all others.   It is by means of these prevalent
winds that the clouds are made to perform their
important task in distributing the waters of the
ocean over the land, a task to which we have already
made allusion, and in this respect the Atlantic is
more uniformly affected than the Pacific.

But if this is the case with regard to the regular
and ordinary currents of air whose velocity may be
calculated on, and whose movements are invariable, we shall find that in occasional outbursts of
violent disturbance in the atmosphere, producing
hurricanes, tornados, cyclones, or by whatever
name frightful storms are known, the Atlantic is
not less terrible than the larger ocean of the
Pacific.

From the West Indies, in the autumn months of
the year, storms of this kind originate from time to

time, and once set in motion, they proceed with vast rapidity in a spiral curve over the sea, occasionally to a distance of from five hundred to one thousand miles. The actual breadth of such hurricanes is not considerable, but they disturb the air and produce ordinary storms for some distance on each side. They appear to originate in the formation of a partial vacuum, towards which, air rushing from all directions, assumes the whirling motion so characteristic of this class of atmospheric disturbances.

Such storms are not confined to tropical latitudes. A remarkable cyclone, or spiral hurricane, passed over a portion of England and the British and St. George's Channel, during the autumn of 1859, destroying an enormous amount of shipping, and where it crossed the land, clearing a way for itself by rooting up trees and tearing down all obstacles. Similar storms are more frequent in the tropical parts of the Atlantic, occurring at intervals of a few years, and they are among the most destructive agencies of nature.

It is curious to find deposited from the currents of air that sweep over the Atlantic occasional showers of dust in a very fine state of division, conveyed in some unknown manner from distant lands, after being lifted high in air to those upper currents of the atmosphere which cross the air currents nearer the surface. This rain-dust has been chiefly observed in spring and autumn, and

most frequently in latitudes between 17° and 25°
north. Its colour varies from pale straw to dusky
red, and it consists for the most part of the
extremely minute skeletons or shells of *Diatoms*
and *Foraminifers* — vegetable and animal forms
which will be described when we come to speak of
the inhabitants of the ocean depths.

These organisms are probably obtained from
some of the great river-valleys of the northern part
of South America, being lifted up in vast clouds of
impalpable sand by the fierce gales induced at the
time of the equinox by the intense heat of the
soil.

"When, under the vertical rays of the tropical
sun, the parched earth crumbles into fine dust, and
the soil cracks as if by the force of an earthquake,
two opposing currents of air producing a rotary
storm come in contact with the soil, a singular
spectacle is seen. The sands rise in funnel-shaped
clouds from the earth, expanding as they mount in
the rarefied atmosphere, and sweeping on like a
waterspout. A dim, yellow light gleams through
the lowering sky, the horizon contracts, and the
wide Steppe seems to draw nearer and nearer to
the traveller. The hot and dusty earth mixed with
the air forms a cloud, shutting out the sky and increasing the stifling heat of the atmosphere. The
east wind evaporates the pools of water before pro-
tected by the leaves of the palm tree, and the large
crocodiles and serpents bury themselves in the dry

mud. During this time, while death and destruction are abroad, the thirsty wanderer is deluded by a phantom of an undulating waterlike surface produced by the deceptive refraction of light." Such, as described by Humboldt in his "Aspects of Nature," when speaking of the plains of Orinoco, is the condition on land which produces the singular dust-rain, falling on the sea at a distance of thousands of miles from where it was lifted.

The icebergs floating on the bosom of the Atlantic, loaded with mud and stones torn away from the frozen lands within the Arctic circle, form a strange contrast to the dust-rain and its source just described. Of such contrasts, however, we meet many in nature, and they have helped each other in producing the world we inhabit.

The fields of ice that float in the Polar Seas are often twenty or thirty miles in diameter and some hundreds of feet in thickness. It is calculated that upwards of 20,000 square miles of drifting ice come down every year along the coast of Greenland into the Atlantic, moving on during winter at the rate of about five or six miles per day. Other equally large masses of ice drift down Baffin's Bay, as much as 2000 miles from their origin. These all melt in the Atlantic, and there deposit whatever solid material had accumulated on them. Some are stranded on the great banks of Newfoundland, others reach much further south, and even cross the Gulf Stream, owing to the great depth of the floating mass, and the strength of the under current setting southwards from the Arctic Sea. Not unfrequently these huge islands advance in large groups into latitudes where they are not expected, and ships crossing from England to America have been caught and destroyed by them. Others have escaped after incurring the most frightful danger. Yet, in spite of all risk, even small yachts have been taken boldly into the most dangerous seas, and have successfully threaded their way through the drifting ice mountains. A wonderful illustration of this was the little "Fox," navigated safely by Captain Sir Frederick M'Clintock, R.N., in his recent successful search after the remains of Sir John Franklin.

Such are some of the phenomena of the Atlantic Ocean, so far as regards its surface and the atmosphere above it. Let us pass on to consider the depths of this vast body of water, and thus learn something further of its nature.

IV.

THE GREAT DEEP AND ITS INHABITANTS.

Interest attaching to the sea bottom—Mode of ascertaining its depth by ordinary soundings—Brooke's sounding apparatus—Massey's modification of Brooke's deep-sea lead—Fitting-out of the Cyclops—Line of deep soundings across the Atlantic—Telegraph plateau as determined by soundings—Mud at the bottom of the Atlantic—Its nature and contents—Animals whose remains occur in the mud.

THERE is something singularly impressive and affecting to the imagination when, in a perfectly calm tropical sea, under a vertical sun, one is able to look down through a depth of forty or fifty fathoms of clear water, and see at the bottom glimpses of a world peopled, we imagine, with all kinds of strange and wonderful objects. There is, indeed, good reason to suppose that these depths lose, in their exemption from surface storms, no small part of their power to support animal or vegetable life; for imagination in this, as in many other cases, has little to do with reality, and the fancied wonders that we strain our eyes to see, if laid bare, would

E

for the most part be found not to differ much from very ordinary objects, and might be little suggestive of beauty or usefulness.   In spite of such considerations, however, no one can be indifferent to an inquiry which has for its object to discover and determine with precision the depth, the form, the nature, and the material of the ocean floor. Such an inquiry has been prosecuted with some success within the last few years, and both the method of obtaining really deep soundings, and the results of them when obtained, are among the novelties of physical geography.

A little reflection will show that to determine, with any approach to accuracy, the depth of water, when that depth approaches or exceeds the elevation of high mountain tops above the sea, is a serious matter.   It must evidently be ascertained with a string or sounding-line; and here, at the outset, difficulties present themselves—for to make such a string strong enough, and yet light enough to answer the purpose, is no easy matter.   Let the reader think of how some seven or eight miles of line are to be got for an experiment of this kind, what it would weigh, where it is to be packed, and how it is to be thrown into and lifted out of the water.   The great difficulty, indeed, is not in getting any required number of fathoms, or even miles, of strong line, or stowing it away.   It is the dropping it into the sea with such a weight that it shall reach the bottom without being floated off indefinitely by under-currents—contriving that the ship itself shall not drift along or continue in motion through the water by wind, by surface currents, or by its own momentum during the time of sinking the line—finding out when the weight strikes the bottom, and therefore when to stop giving out line in consequence of the operation being so far completed—these are among the difficulties to be overcome before reaching the bottom. And then, when the line and lead are down and the bottom is reached, we have to bring the line back into the ship, and, with it, we ought to have some proof of the nature of the sea bottom.

Easy as these things seem to be, and are, at moderate depths, they have required all the resources of modern science, and all the ingenuity that could be brought to bear upon them, before they were found manageable in those deeper parts of the ocean called by sailors " blue water."

Everybody who has been at sea has probably noticed the operation of sounding, or throwing the lead, when the vessel approaches land, and it is uncertain what may be the exact depth of water in the line of her course. In this operation the lead has a cup-like hollow on the lower surface, to which a lump of tallow is attached, and the loose particles of mud, sand, or shell, if there are any, adhere to the tallow, and are brought up with it, thus showing not only the depth of the water by the length of line run out, but the nature of the bottom.

If there is nothing brought up, the inference is that the bottom is rocky, and sometimes the impression of the jagged surface is seen. It has been found that this method is only adapted to small depths (within 600 feet); but certain modifications have long been in use by which depths up to two or three thousand feet could be determined, at least approximately, although, when this quantity of line was run out there has never been felt any certainty that the weight really reached the bottom; or, if so, that a great length of line had not been wasted in forming a curve, owing to the action of submarine currents dragging the line. In some cases, as much as 50,000 feet of line have been run out without any proof being obtained that a bottom was reached; and owing to these failures, many parts of the ocean have been thought unfathomable.

We have to thank our brethren from the other side of the Atlantic for a number of trials and experiments, with various modifications of the old sounding-line, and also for the introduction of a simple and efficacious contrivance for overcoming the difficulty. Brooke's sounding apparatus, slightly modified in matter of detail, is now generally employed, with the greatest success, to obtain proof not only of the depth, but of the nature of the bottom of oceans, even where the distance to be traversed is greater than the height of the loftiest peak of the Himalayans or the Andes above the sea level.

A description of the apparatus, and of the modified form of it employed in the British navy, will not be without interest.

The principle involved in this contrivance is very simple, and the manner of working it out is equally ingenious. The practical difficulty has been to send down a line sufficiently weighted to sink rapidly to the bottom, and sufficiently strong and small to be lifted back again in moderate time without injury. Mr. Brooke contrived an apparatus which, in itself, was a light frame-work, containing a cup and valve for catching and holding the mud or sand of the bottom, but to this was attached a heavy sinker, in such a way that, while perfectly safe to carry down the line, it became detached, and was got rid of the instant the bottom was reached. There was then nothing to bring up but the line itself, and the few pounds of framework, with its mud. In this way the bottom was reached with certainty, and nothing was lost but the sinker, which might be an iron shot of sufficient weight for the purpose.

It may perhaps be said why, if the line did not break with the weight of the sinker in going down, it should not lift it with equal safety; but a very little explanation will make this clear. When an object is dragged through water, even near the surface, a considerable resistance is experienced, owing to the friction of the water; but when a line with a weight attached is sunk in deep water, the pressure of the water on the surface exposed, increasing with the depth, soon begins to produce an effect; and from being very small within a few scores of fathoms, where the pressure is hardly felt, becomes enormously great at a depth of several hundred fathoms. Thus at a depth of 2400 fathoms (14,400 feet) the pressure of the water is as much as three tons on every square inch of surface; and since 2400 fathoms of ordinary whale line, such as is used in deep soundings, would weigh about a ton, and would present a surface of one square foot for each fathom, or 2400 square feet in all, the friction of moving the line up through the water at this depth would be enormously great. The additional friction of a large weight at the end of a line exposing a considerable surface, would indeed be exceedingly more difficult to overcome than the mere weight itself would seem to justify. It was found on board the *Cyclops* that, in order to haul in a line of 2400 fathoms without the sinker, it was necessary not only to use a twelve-horse power steam engine provided for the purpose, but that the steam had to be raised so as to obtain a pressure of twelve pounds on the square inch before the inertia could be overcome and the line set in motion through the water. In this case the actual weight lifted did not exceed one

ton. The strain on every part, but especially on the
upper part of the line, during this pull must be
enormous, and no one will wonder that in many
cases the line has parted and been lost during
the operation. It will also be understood how and
why it was that, before the sounding apparatus
was used, and the sinker detached, no result of any
value could be obtained; for either the line was too
weak and broke at the first attempt to lift it with
the weight attached, or the line was so large and
the weight so small that the bottom was never
reached. It will serve to give a further idea of
the pressure to record that, in one of the cases referred
to above, which took place on board the *Cyclops*,
"the tar was forced out of the rope in an extraordinary manner, several of the splices started, and
the rope was much stretched."

To come back, now, to the construction of the
sounding-apparatus. The original contrivance of
Lieutenant Brooke consisted of a cannon shot,
having a hole through it for the passage of an iron
rod, terminated upwards by a pair of hooks, from
which the shot was so slung that the ball was
detached when the bottom of the sea was struck,
and the rod, with a contrivance at the bottom
to receive the mud, was relieved from its
weight and could be lifted. A more elaborate
machine, on the same principle, was prepared for
use in the British navy, and is thus arranged:—

Fig. 1.   Fig. 2.   Fig. 3.

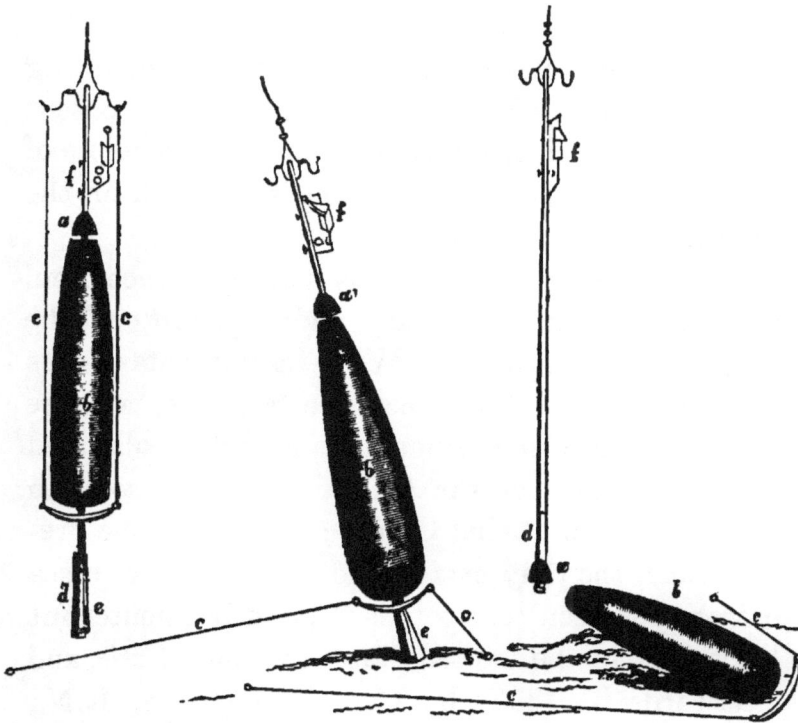

Modification of Brooke's Sounding Apparatus in use in the British Navy.

Fig. 1 represents the apparatus ready for letting go; fig. 2 its condition at the moment of striking the bottom; and fig. 3 shows the rod while being pulled up, after it has struck the bottom, and got rid of its heavy sinking weight, and received the sample of the bottom mud which is to be brought up. In these figures, *a* represents the small leaden weight; *b* the heavy iron sinker; *c* a pair of rods attached to a saucer, or plate, which supports the sinker, and is suspended from the top of the apparatus; *d* is a valve adapted to sink into and receive the contents of the bottom; *e* is a small spring, by the action of which the weight *a* descends and closes the mouth of *d*, remaining afterwards at the bottom; and *f* is Massey's patent sounding machine, constructed on the principle of Massey's patent log, and serving as a check on the quantity of line run out.

The action of the apparatus will be at once seen, and the extreme simplicity and ingenuity of the contrivance recognised. With instruments of this kind, and in several successive voyages, a large number of deep soundings have now been obtained in various parts of the Atlantic, and also some in the Indian and Pacific Oceans; but up to the present time, the only systematic exploration of a sea bottom has been that conducted first by Lieutenant Berryman, in the United States steamer *Arctic*, and afterwards in 1857, by Captain Dayman, R.N., commanding the British steamer *Cyclops*. Both investigations had the same object, namely, to discover the depth of the Atlantic in the line on which it was intended to deposit the cable for the Transatlantic electric telegraph.

The *Cyclops* was especially fitted for the work she had to perform. In addition to some 27,000 fathoms (about 30 miles) of line of different kinds, and eighty self-detaching iron weights to carry down the line, she had a special steam-engine placed so as to heave in the line, and an ingenious arrangement enabling the ship to be kept in one spot on the sea during the time the lead was sink-

ing, and this not only in fine, calm weather, but
even in a fresh breeze, with a high sea. It may be
assumed that no surveying ship will be sent out
again without similar means of ascertaining the
depth of the ocean in various latitudes and in every
sea.

The actual positive results of deep sea soundings
across the Atlantic, and in different parts of its bed,
are neither few nor unimportant. They lay bare,
as it were, the general features of that large, depressed tract, extending between Europe and Africa
on the east, and the two Americas on the west, and
show, in some measure, how far the general features
of the land, as seen in the bounding continents,
extend and influence the space between them.

The line selected for soundings had reference to
the intended position of the Atlantic telegraph.
The American soundings had appeared to show the
existence of a kind of plateau having no marked
differences of level for the greater part of the way
across between Ireland and Newfoundland. The
soundings made by the *Cyclops* confirmed this view.
Starting from Valentia, casts were taken, on an
average, every fifteen miles, for a distance of two
hundred and fifty miles. Up to this point the
depth was inconsiderable, the water deepening
gradually, with a sandy bottom, till about 500 feet,
and then more rapidly, but still not more than with
a gentle slope, till, at a distance of 120 miles from
land, the depth was found to be 2500 feet, with a

bottom of hard rock. This only gives an average slope of twenty feet in a mile, and if an even slope it would be imperceptible, though no doubt being accompanied with undulations, it admits of a fair amount of hill and dale. For the next hundred miles there is a little more change of level, though still not amounting to more than a gentle sweep rising to the west, and this is followed, still going west, by other slopes, first downwards but chiefly upwards, till, at about 230 miles from the coast of Ireland, the water is only 1320 feet deep. So far, therefore, the general features of the land are preserved; and if the whole ocean floor were lifted up 3500 feet above its present level, or the ocean level itself were to be diminished by so much, the general features of the land in western Europe would be in no respect interfered with. A wide, low plain would extend from the new coast line to the cliffs which now bound Ireland—small but unimportant depressions would mark the site of the British Channel, the St. George's Channel, and the German Ocean, and the mountains of Wales and Scotland would rise between 7000 and 8000 feet above the new sea level. There is reason to believe that this general similarity of the form of the land covered by the sea to that now above its level, would continue down the whole western coast of Europe.

After this point is reached, however, a change takes place, and the true Atlantic depression commences. Within the short distance of twenty miles, the difference in soundings marks a sudden fall of the sea bottom of upwards of nine thousand feet. Here then is the first deep sounding, and from this point the real novelty of the subject commences.

The reader will have observed, if he has followed the account just given, that there is no indication whatever of any alteration of level till this depression is reached. It is not a chain of submarine mountains that is come upon, for the sea bottom has been comparatively shallow for the whole distance from the Irish coast, and consists apparently of gentle undulations, precisely similar to those of the greater part of the land. But from the edge, as it were, of a great plateau, the ground suddenly drops, the depression being fully equivalent to that of the Alps on the Italian side, and very much greater than that of the Pyrenees from the high plateau of Spain towards France. The depression, therefore, must be regarded as a true marine cliff—a descent from a plateau at one elevation either to another plateau at a lower level, or into a deep gorge.

We are at once carried from moderately deep water into the great depths of the ocean, and as from this point, for a distance of twelve hundred miles, there does not seem to be a single exception to the general level condition of the ocean floor, it is quite clear that we descend not into a gorge, but to a new, lower plateau, a country hitherto altogether unknown.

We describe the condition of the plateau as generally level, and this is really the case, although there exist differences of elevation which at first sight appear considerable. The extreme depth of the lowest point of the Atlantic, below the foot of the cliff just described and on the line of the proposed telegraph cable, is a little more than 4000 feet, and only in one spot of the many on which soundings were taken was there any elevation 1500 feet above the level of the cliff bottom. For at least 1000 miles it does not appear that there is a difference of level between the deepest and loftiest point attained amounting to 5000 feet; so that, in fact, the whole surface is one vast depressed plateau, totally unlike any equal extent of dry land, though more resembling that on the eastern side of the Andes, in South America, than any other known land. On the American side of our Atlantic plateau there is a second cliff, facing eastward, having a total rise of about 5000 feet, immediately to the west of which the ground slopes gradually upwards at the rate of about forty feet in a mile, till it reaches the American continent.

This plateau is a novelty in physical geography. On a small scale, the elevated plains of the earth present the converse of such phenomena; and inland cliffs of a few hundred feet, sloping gradually away from the top to the plain country beyond, are objects of great interest in Spain and elsewhere.

But the proportions in the bed of the Atlantic are gigantesque, and are suggestive in many ways in reference to the geological structure of the globe.

We must pass on, however, to the condition of other parts of the Atlantic floor. The deepest part of the North Atlantic appears to be on the American side, some distance south of the great banks of Newfoundland, between the 40th and 35th parallels of latitude.

There seems to be here a basin, whose axis ranges east and west, having nearly a thousand miles of length, and whose depth below the present sea level is greater than the elevation of the highest mountains in the world above that level.

Whether this great amphitheatre is approached by cliffs, forming, as it were, a succession of terraces or steps to its deep floor, is not yet clearly determined; but there appears little resemblance in the form of the hollow to an inverted mountain chain, the area of the deepest portions being far larger in proportion, and the scale altogether more considerable. Proceeding from the lowest depths, as here determined—depths more than 30,000 feet (about six miles) below the mean level of the ocean —the bottom seems to rise gradually to the north; and there, at 10,000 feet, it forms the terrace or plateau already described, which is supposed to occupy not less than a million of square miles. At the south corner is the coralline group of the Bermudas rapidly approaching the surface, and separated from the West Indian islands by a trench of some 20,000 feet, deepening towards the north-east coast of South America.   Towards the East the depression appears to be somewhat less considerable, but always very great, till we see the islands of the Azores rise suddenly out in the form of volcanic cones, to be again separated from the shores of Africa and Europe by a depth of 15,000 to 18,000 feet of water.   The Cape de Verd islands also rise very steeply from exceedingly great depths, like mountain peaks, shooting up to enormous elevation from a comparatively low range.

Such is the state of knowledge with regard to the form of the Atlantic ocean floor.   We cannot yet, indeed, map out in detail its physical features, or mark down all its valleys and ridges, its mountains, level plains, and rolling or swelling prairies, the exact line of its cliffs, or the precise bearing of the deep, narrow gorges that intersect it; but much has been done in the way of making out the general outline, and it only remains for future observations to fill in the contour lines.   A very important step has been made in the progress of discovery, and a new field of accurate observation presented for investigation.

The deep soundings that have been made in the Pacific are too few to justify any generalizations whatever.   So vast a space has there to be surveyed that it will probably be some time before we can know, even by the most rough approximation,

how far the form of its bottom corresponds to that
of the Atlantic, or whether it agrees more with that
of the land now above the sea. The limits of depth
in this ocean are also, at present, undetermined.

While observing the depth of the ocean, observations have been from time to time made as to
the temperature of deep water. It was known,
from experiments made nearly a quarter of a
century ago, by General (then Captain) Sabine,
in water, under the tropics, supposed to be 6000
feet deep, that the temperature, at the lowest
depth reached, was only 45·8°, that of the surface
being 83°. The careful result of thermometer
observations, made with every precaution by the
officers of the *Cyclops,* shows that in temperate and
warm latitudes the water is uniformly colder at
great depths than at the surface—that it generally
shows a gradual increase of cold in going down,
and that at depths of 10,000 feet within the tropics
the temperature is tolerably equable at 40° to 46°
Fahrenheit, and is altogether independent of the
heat of the water at the surface.

The nature of the bottom, whether sand, rock,
or mud, is a matter quite within the range of actual
observation in sounding with the deep sea apparatus,
since the quantity of soft material brought up, if
any is present, is sufficiently large to afford definite
results. It is not a little remarkable that, with
two exceptions, *all* the soundings made over the
10,000 feet plateau across the Atlantic, and most

of those made elsewhere, have brought up a material precisely similar in every case—described as a soft, mealy, sticky substance, light coloured and mudlike, called by the surveyors "*oaze*." On minute examination, this was found to be an impalpable powder, with a mixture of slight grittiness, the powder composed of carbonate of lime, nine-tenths by weight of the whole mass consisting of very minute animal organisms, and the gritty particles (not much more than five per cent. by weight of the whole) being angular fragments of some hard mineral. The remaining small proportion includes flinty skeletons of animals and vegetables. These results in themselves are singular enough, but it is still more remarkable that everywhere these animal organisms seem to have belonged almost exclusively to varieties of one species, whose microscopic skeleton has been distributed in inconceivable numbers over a million of square miles of ocean floor.

The animals to which these skeletons once belonged are referred to a group showing the smallest amount of organization. Simple, however, as they are in point of mechanical contrivance, the great mystery of life is not less wonderful in them than in the structure of the largest quadruped, or of man himself, and the little deposits of carbonate of lime they accumulate, to enclose or cover the bag or cell of jellylike matter of which they are composed, is as regular and as well adapted to the creature as any skeleton, carapace, or shell ever

F

constructed. Conceive a creature, thousands of whose fellows might disport themselves with ease in the drop of water that would rest on a small pin's head, a mere lump of jelly, capable of assuming any form whatever, and actually doubling itself and becoming a temporary stomach whenever an atom of food happens to present itself; imagine this creature vaguely extending filaments, like very fine rootlets, in every direction, and drawing them back when it pleases to form with them parts of a new stomach; conceive also a number of these combining together, each acting the part of a separate individual, but all having one common substance—each, notwithstanding its extreme simplicity of structure, endowed with the power of separating particles of carbonate of lime or silica from sea-water—a faculty with which not every chemist is gifted—and building with this carbonate of lime or silica a kind of coat, a *carapace*, or skeleton. In a group of such individuals all are so closely connected as to form apparently one mass, but each builds for himself a construction of the same kind, until at length there is obtained a complicated, many-chambered shell, not much unlike that of a nautilus, only infinitely smaller, which remains a permanent monument, enduring for ages, long after the larger races have died out and been destroyed. Such animals may, as they do now at the bottom of the Atlantic, afford the only proofs of animal existence over areas thousands of

square miles in extent, although in the waters above
there may have floated innumerable fishes and
other animals far higher in the scale of existence,
and for this very reason far less able to resist the
accidents of fortune.

Fig. 4.

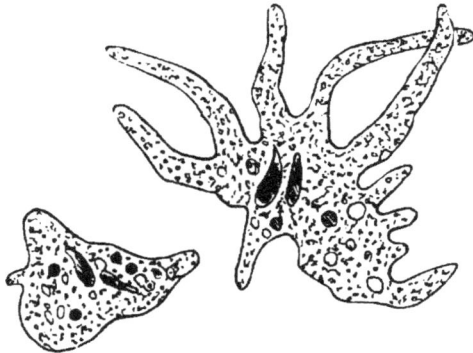

AMŒBA.

The annexed wood-cut will illustrate the varieties of form and appearance of the simplest of
these singular creatures. It represents (see fig. 4)
a mass of jellylike substance, full of small irregularly distributed bladders, and the two figures represent only two of the infinite variety of forms
into which it is capable of throwing itself. When
this creature meets with a particle of nutriment, it
simply encloses and digests it, throwing away any
undigested part. It is not possible to imagine a
state of existence more simple.

The first step in advance is made by these

animals when they secrete a strong covering, and throw out threadlike fibres to collect food. In figure 5 *a* the soft animal is represented throwing

Fig. 5.

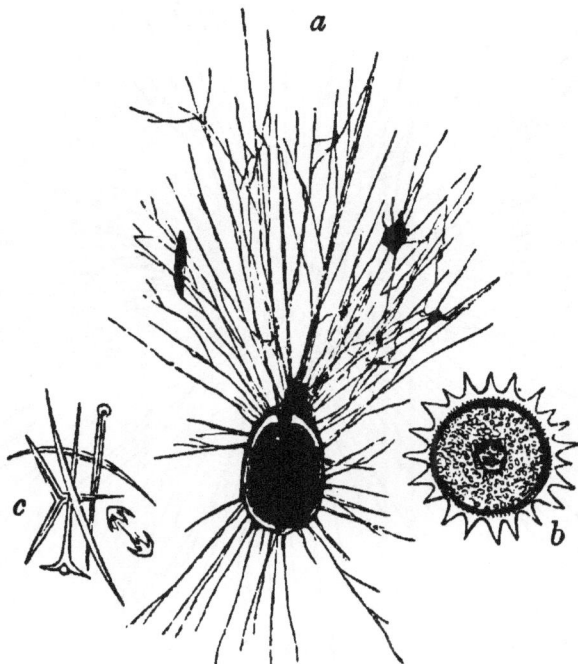

a. Rhizopodous animal.   *b. Polycystina.*   *c. Sponge spicules.*

out its fibres with which food is entangled and conveyed to the central mass. In this animal there is a simple strong covering, and no complication of any kind is observable; but in the next, figure 6, is seen the compound shell, which, however, is merely a repetition of the simpler one. The shells which form so large a part of the deep Atlantic mud are of this latter kind, and belonged to an animal exactly similar.

Fig. 6.

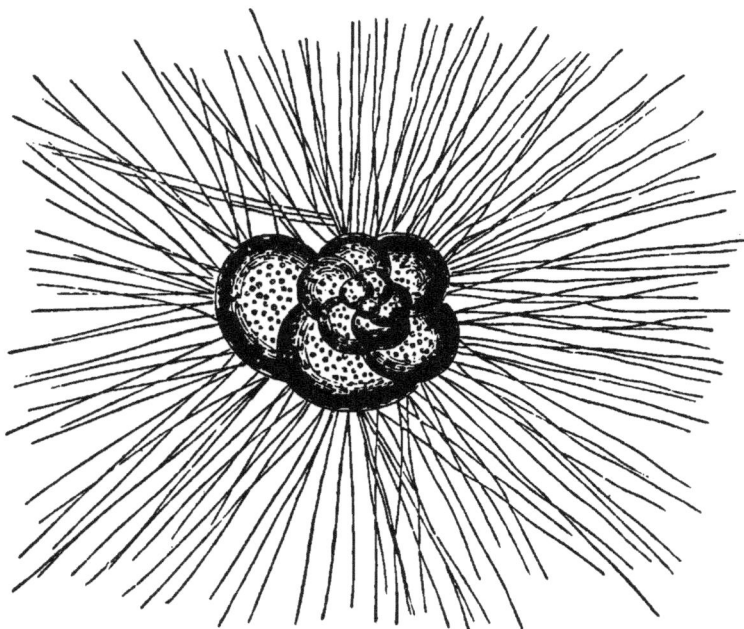

RHIZOPOD.

View of the compound shell and animal, with the filamentous processes.

It might naturally be expected that a compound animal, consisting of a multitude of independent existences connected in one body, would not always secrete shells of exactly the same form. No doubt there has been a long succession of such creatures, and they have changed with changing circumstances—no doubt, also, many of those now living will be found to have lived through long geological periods. In this earliest form of animal life variety of detail appears to be an absolute necessity of nature.

Fig. 7.

Broken fragments of GLOBIGERINA, *as seen under the microscope, in the ooze from the bottom of the Atlantic.*

There is one form, however (fig. 7), which, much more than any other, crowds into view whenever the Atlantic mud is examined. The individuals are sometimes of comparatively large size, but, whether large or small, whole or fractured, the fragments of their shells abound everywhere. It is probable that conditions of existence especially favourable for the development of this form may be the chief reason of its occurring so plentifully.

Besides the calcareous skeletons, which form, as we have seen, nine-tenths by weight of the mud, there is another group of even more minute animals, who separate from the water flinty, instead of cal-

careous, valves or cases, and these flinty skeletons
are also very widely spread. The forms assumed are
singular and fantastic in the extreme, and they are
often extremely beautiful objects under the microscope (fig. 5, *b*). Notwithstanding their small size,
important deposits of them are found in many parts
of the ocean.

Sponges are animal substances composed of a
fibrous net-work, strengthened by needle or anchor-
shaped spines of flint or limestone, and clothed with
a soft flesh provided with hairlike filaments lining
cavities in this soft coating. Through the canals
which these cavities form, currents of water are
kept continually passing by the vibration of the
filaments, bringing in food and removing the undigested and unassimilated particles. A few of the
spines or spicules of flint or other stone (fig. 5 *c*),
complete the list of animal structures found in the
mud at the bottom of the Atlantic.

No doubt the reader will be astonished that no
remains of animals of higher organization, living
in the water above, have yet been found with this
mud and sand, but the very depth of the bottom is
a reason why such remains never find their way
thither. After death, the carcases of the larger
marine animals become the prey of others of smaller
size and lower organization, and these again in
their turn are the prey of others, till at length we
reach the minute organisms described.

In some places the thickness of the chalky

mud thus deposited, appears to be but small, and in one or two soundings the mud was replaced by shells, or even pebbles. Additional observations may bring to light other facts and increase our knowledge of this curious history, but hitherto the fine mud loaded with, or rather made up of, organic bodies, has proved to be the most abundant material at the bottom of deep water, while sand and rock are more common in the shallower parts of the Atlantic.

Fig. 8.

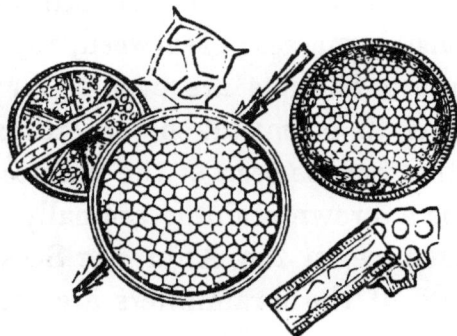

Group of envelopes of DIATOMACEÆ.

Lastly, we find mixed up with this fine mud, a small proportion of the flinty envelopes enclosing some of the very simplest forms of vegetable life. These early stages of vegetation resemble very much the early developments of animal life, and are even endowed with spontaneous movement, although apparently without organs of locomotion. They consist of simple cells, extending rapidly, and multiplying by a natural division of one cell into two, but each cell is coated with a thin film of flint, secreted from the water, and often exhibiting the most exquisite beauty of form (see fig. 8).

They increase with a rapidity that seems almost miraculous; and, minute as they are, they sometimes choke up harbours and diminish the depth of channels. One deposit of their valves, or films of flint, is mentioned by Dr. Hooker as being not less than four hundred miles long and one hundred and twenty miles broad, and of great and constantly increasing thickness. It must not be supposed that these vegetations are ultimately developed into visible masses of sea-weed, any more than that the minute animals are ever so far magnified by growth, as to be mistaken for ordinary shells. The most that can be seen of the minute vegetable organisms is a brown stain occasionally noticed on newly formed ice in the Antarctic Seas; and the largest shells of the foraminifers are rarely larger than a pin's head, most of them being inconceivably smaller, and not visible to the unassisted eye.

In some parts of the Atlantic, where coral is abundant, the sea bottom is covered with a different and more brilliant show of animal life; but corals, although not unimportant in the tropics, are comparatively small in the temperate and colder seas. They do, however, exhibit beautiful varieties of colour, and occupy extremely large tracts.

The living coral is chiefly to be found on the exterior line of an extending reef, the interval between which and the land (or the interior space where the bank or reef forms a detached islet) is occupied by broken fragments and smaller kinds of similar animals. There is a dull unsightly sort reaching to the depth of 400 fathoms; but with the exception of a few kinds, none are found much below fifty fathoms, and those which chiefly build are limited to twenty fathoms.

Over all the great plateau, then, which has been described, and throughout those successive terraces that range deeper and deeper towards the lowest point of the Atlantic floor, we have no proof of the existence of any substance accumulated but the fragments of the minute and lowly organized specimens of life just described.

Strange absence of what might have been anticipated, or of what the imagination would have pictured, as likely to be present under the circumstances. What has become of the innumerable bones and teeth and scales of fishes that, for all the years gone by, have died in the broad Atlantic? Where are the remains of the many ships that have been swallowed up by its waves? Where the gravel heaps left behind by the icebergs that have been melted in floating down from the Polar Seas? Where, also, the substances drifted across by the Gulf Stream and other currents that traverse the ocean? Nothing—not one solitary indication of all these, but in their place a fine, impalpable, tenacious mud, everywhere extending, and made up of little particles of carbonate of lime secreted by countless myriads of animalcules, the food perhaps of the whales and fishes of the surface, but more probably the sole inhabitants of those great depths which other animals more highly organized would in vain attempt to penetrate. Truly may we say that the secrets of the great deep are mysterious and grand—and that the search after them amply repays the labour of investigation.

By the investigations of modern observers we find laid open for our observation the great valley of the Atlantic, descending by a succession of broad terraces and steplike cliffs, to a gully or ravine some 70,000 feet below the high ridges of the Andes and Himalaya Mountains. Like a vast amphitheatre, these terraces appear to surround a central comparatively small arena. On one principal upper floor, ten to fifteen thousand feet below the present sea level, we have discovered the inhabitants, and already all trace of our world above is lost—all high forms of organic life are absent, and nothing remains but skeletons of the simplest animals, of which it may almost be said that but one or two specific forms can be determined—certainly nine-tenths of the solid material consisting of the remains of one species only. What may be expected when the deeper levels are dredged and we bring up the material of which they consist? What will be the direction of the line of deepest depression, and how will it agree with the adjacent lines of greatest elevation? What will be the inhabitants of that part of the ocean bottom, fully double the depth concerning which we have only very recently obtained information, where the pressure of the water is nearly seven tons to the square inch of surface—and where each cubic foot of water probably occupies less space by one hundred cubic inches than it would do if lifted to the surface, owing to the weight of the superincumbent column of water resting upon it.

These are queries yet unanswered, but to many, if not all of them, we may expect to receive satisfactory answers before very long. The facts that remain to be determined are hardly more obscure or difficult to make out than those we have already mastered, and the investigations when once fairly set on foot, will probably soon leave little to be desired. The soundings that formerly could not be attempted, except in the calmest weather and with the certainty of half a day's severe and incessant labour, are now completed, not only in ordinary weather, but even in a fresh breeze and disturbed sea in a couple of hours; and whilst, according to former methods, there was more probability of enormous error than of accuracy, the result is now to be depended on in almost every case within very moderate limits.

V.

THE INTERIOR OF AFRICA.

State of knowledge of the interior of Africa in 1850—Cause of the ignorance that prevailed—Travels across the Great Northern Desert, or Sahara—Dr. Livingstone's researches—Discovery of Lake Ngami—Water system of the Upper Zambesi—Country between Lake Ngami and Loanda—Character of the coast range—Descent of the Zambesi—Exploration of Burton and Speke from Zanzibar—Coast range—Discovery of the great Lakes—Probable source of the White Nile and absence of the Mountains of the Moon as a high range near the Equator—General résumé of the physical geography of Africa and its geological structure.

From the first voyage of Bruce to Abyssinia in 1769, up to the close of the year 1849, a period of eighty years, there had been upwards of twenty important expeditions into the interior of Africa by different European travellers, chiefly English and German. With one exception—that of Lacerda in 1796*—all these had reference to the northern

---

* Lacerda was a Portuguese, and there is reason to believe that some of the more adventurous of the Portuguese had penetrated far into the interior and become acquainted with the general outline of the physical geography of Eastern Africa before the attention of English and German travellers was directed to the subject. Since, however, the information they may have

part of the continent, and except in the case of
some attempts, commencing on the Guinea coast,
had not succeeded in reaching within ten degrees of the equator.   Of the greater and more
determined efforts to penetrate into the interior,
almost all commenced from the north; but owing
to the nature of the country, and the extreme
difficulty of crossing the Great Desert, the points of
departure were a thousand miles asunder, and in
fact were limited to three positions—Cairo, Tripoli,
and Tangier.   From Cairo, Bruce, and afterwards
Browne and Burckhardt, working up the course of
the Nile towards its sources, succeeded at length
in determining the main tributaries of that important stream with some precision.   From Cairo, also,
Hornemann, in 1799, made his way along the
northern part of the Great Sahara to intersect what
may apparently be called the great high road of
native traffic from Tripoli to Timbuctu.   From
Tripoli, starting in 1821, Major Denham and Captain
Clapperton made a successful trip across the Great
Sahara to Lake Tsad, and thence, after exploring in
various directions, reached Sakatu.   Afterwards,
in 1826, Captain Clapperton advanced to Sakatu
from the Guinea Coast, thus completely crossing
Northern Africa, from sea to sea, on a line measur-

---

acquired never was communicated to the world, and probably never
would have been but for its re-discovery, it does not appear that
much credit is due to them. They drew no scientific conclusions,
and applied what knowledge they had to no useful purpose.

ing more than two thousand miles as the bird flies. Lastly, in 1828, Caillie succeeded in crossing southwards from Tangier to Timbuctu, and thence traversing the country to the west, he reached the Atlantic coast at the mouth of the Gambia. It is known that Mungo Park, in 1805, had advanced from the mouth of the Gambia as far as Timbuctu, and he seems to have descended the Joliba, the main arm of the Niger, as far as Boussa, from which point its course is known. The death of that unfortunate traveller, however, and the loss of his notes, has prevented an exact record of his course from being preserved.

The state of African geography, then, at the period alluded to, was in the highest degree obscure and unsatisfactory, notwithstanding the exertions of so many able and energetic travellers, most of whom lost their lives in endeavouring to bring home more complete and satisfactory accounts. Through the vast tract north of the equator, containing about six millions of square miles of country, extending to the thirty-seventh degree of north latitude, and from 10° west to 50° east longitude, there had been but three traverses made from the north towards the equator, and one from the west coast, near the sea, to a point about mid distance across. Something also was known of the eastern coast on the same parallel. There still remained, however, a gap nearly a thousand miles wide between Lake Tsad and the White or Western

Nile, of which it may fairly be said that nothing whatever was known either by direct observation or the statements of native travellers. The ignorance of Central Africa south of the Equator extended to the whole country as far as the Orange River, in latitude 25° south, within a few days' journey of the colonies of the Cape of Good Hope. Several travellers had advanced some distance from the Cape Colonies, but except from the journey of Lacerda up the Zambesi in 1796, and a sporting excursion of Gordon Cumming in 1843 and 1848, which latter led to no important geographical discovery, little had been done within these limits, which reach from the Mediterranean to within a few hundred miles of the Cape of Good Hope.

Nor was this all. The width of this vast tract of fifty-six degrees of latitude—nearly four thousand miles of unexplored country—was in many parts a thousand, and in some more than fifteen hundred miles, only an extremely narrow strip of coast on each side having been visited, and within this strip no white man had been known to penetrate. From the Portuguese settlements on the west to those on the east coast, there was a complete blank, and the prevailing opinion seemed to be that as the intervening country had not been visited, it was not habitable. The names of a few rivers were recognised, but their magnitude and direction were altogether unknown.

Such, ten short years ago, was the state, not of

knowledge, but of ignorance, concerning a tract of
land occupied by the human race perhaps longer
than any other; rich as we now know in vegetable
and mineral wealth, crowded with animal life, not
unprovided with rivers or lakes—not inaccessible
by lofty mountains cutting off communication—
not more unhealthy, in all probability, than many
other countries much better known, and yet owing
to a number of causes by no means easily described,
the most obstinate of all in resisting the advance of
civilization.

For it must not be supposed that there has been
any want of adventure on the part of African
travellers. Nowhere, not even in the Arctic Ocean,
have men more perseveringly braved every kind of
personal suffering and death itself, than in the vain
attempt to penetrate into the interior of this closed
continent. Frustrated in one direction, they have
again and again endeavoured to find another less
difficult of access. Now entering the dreary
and hopeless wastes of the Sahara, now in boats on
navigable rivers—now landing in the fatal swamps
of the tropical rivers of the Guinea Coast—some
travelling as Arabs and mixing up with the traders
of the country, or taking presents to gratify the
vanity or cupidity of the natives—others armed
with authority and escorts. Occasionally the adven-
turer would start alone, or with only a native
servant, to elude observation; some would go in
companies of two or three or more, to communicate

G

with one another and help each other in time of need. Some were well armed, and some without arms at all—there is, in fact, hardly a mode of travel that has not been tried in the effort to advance into the interior of Africa. And after all how small was the result. It would seem that up to a certain point in discovery of all kinds the path is difficult, tedious, and obscure; but after that point is once reached, the road becomes wider, smoother, and easier—we lose sight of the difficulties which the early pioneers had to go through, and in the rapid movement onwards, we sometimes forget how much we owe to those who led the way.

The advance that has been made towards a knowledge of the interior of Africa, within the ten years since 1849, has been so very great, and has tended so much to connect the various threads of information previously floating uselessly on the surface of descriptive geography, that although large gaps still remain to be filled up, there is no such continuous space of untrodden land as we have just been describing. The labours of Burton and Speke as successive explorers in the north, and of Livingstone in the south, would alone have been sufficient to indicate a probable outline of the physical geography of the continent, which cannot fail to have a most favourable effect on future discoveries; while the latter, in showing the capabilities of Southern, if not Central, Africa for supplying vegetable products of which we are greatly in need, and taking in exchange manufactures which it is the interest of our people at home to provide, has offered an inducement stronger than any other to penetrate yet further and obtain yet larger results.

The country opened out to us by the travels of Dr. Livingstone, extends northwards from the Bichuana country (already known by previous descriptions and traversed by the north-easternmost feeder of the Orange river), to the Portuguese settlement of Loanda, on the west coast, in south latitude $8\frac{1}{2}°$, and to the mouth of the Zambesi, a river whose name indeed was before known, and which will be seen marked in most maps as debouching in latitude about 18° south, but of whose course and embranchment geographers were perfectly ignorant.

In an irregular line between these latitudes Africa has now been completely crossed, and before crossing Dr. Livingstone proceeded northwards through a country also almost entirely unknown, extending from latitude 25° south to 18° south, a distance in direct line of nearly five hundred miles, obtaining as he went a vast amount of information. The map of Africa is now therefore pretty well filled up south of a diagonal line drawn from the fifth parallel on the west side to the tenth on the east, and we have a knowledge of most of this large tract of country, previously unknown, owing almost entirely to the successful labours and investigations of one man.

The Kalahari Desert and the Bichuana country may be regarded as the starting point of discovery. The so-called desert is flat and has no running water, although it is not without traces of river courses and possesses wells with a little water. It is covered with vegetation, and roamed over by prodigious herds of antelopes and their enemies, the larger carnivora, besides many human inhabitants. These latter are partly bushmen—a purely hunting race—and partly the descendants of the tribes occupying the adjacent country to the east, whose habits are those of herdsmen and agriculturists.

After crossing from the Bichuana country along the eastern boundary of the desert to a district of salt efflorescence, a stream is reached, which in time conducts the traveller to the Lake Ngami, exaggerated accounts of which had previously been received by many travellers, and which proved a turning-point in the progress of discovery. Before reaching the Lake, and after following a beautifully wooded river for nearly one hundred miles, always approaching the Lake but still far from it, Dr. Livingstone and his party discovered a large stream flowing into the river, and on making inquiry, learned that this stream came from a country full of rivers, so many that no one can tell their number, and full of large trees. Such was the welcome announcement on the approach to a district up to that time assumed to be a sandy naked plateau.

In due time the river was found to open out into the lake, which appeared to be a shallow expanse receiving several streams, and no doubt altering greatly in magnitude at different seasons. It did not appear, when visited by Dr. Livingstone, to be more than sixty-five miles long by twelve to fourteen miles broad, and its level, as determined by the boiling point of water, is less than two thousand feet above the sea. It appears to connect itself with the waters of the Zambesi, which will be afterwards described flowing outwards at the eastern end by the river Zouga, and it receives from the north-west the waters of a large stream.

A great part of the country between Lake Ngami and the nearest western coast was travelled over in 1850 by Mr. Galton, and the Lake itself was reached by Mr. Anderson from the west in 1852. Mr. Galton found a considerable range of limestone rock, and a country opening out into the interior, intersected by many watercourses; and this traveller, with his companion and successor, in a journey of seventeen hundred miles, succeeded in clearing up all obscurity as to the whole district extending from the coast to the Kalahari Desert. The hilly and even mountainous character of the coast did not seem to extend far into the interior, where the land may be regarded as a plateau, the height of which is not very considerable, although many parts of the coast range are four thousand to six thousand feet above the sea.

To the north-west of Lake Ngami, the stream
flowing into that sheet of water has been ascended
a hundred and fifty miles, and lofty mountains with
white summits are described by the natives as seen in
that direction, marking a considerable elevation of
the coast range.*   Towards the north-east it appears
that there is a junction with another river of considerable magnitude connected with a large but
little known tract north of Damara, hitherto unvisited, but believed to be extremely well watered
and rich, although flat and marshy, covered with
grass, and in places subject to malaria fever.

Dr. Livingstone reached the outfall of the river
proceeding from Lake Ngami, and found himself
then on what he regarded as the main stream
of the Zambesi.   He found the river there more
than a mile broad, with many islands abounding
with vegetation.   The water being high in the
season of his visit (the winter) he was enabled to
pass rocks and rapids that would be a serious
impediment in summer, since in many cases, even
with the stream full, the canoes had to be carried
from one point to another by land.   Numerous
villages were passed, some indeed of large size,
but all peopled, and the country seemed to some
extent cultivated.   For nearly three hundred miles
the stream was followed, large tributaries coming

---

* It appears that the colour of at least some of these mountains, which has been taken for snow, is due to the presence of a
peculiar white lichen covering the naked rock.

into it from both sides, except where at intervals
the hills closed in and formed a narrow and picturesque gorge. At length our traveller reached a
point where the forest coming to the water rendered it difficult to proceed further, and where also
the presence of a remarkable fly (the *Tsetse*) is fatal
to the oxen. Beyond this point our traveller proceeded only to the junction of another stream from
the north at a point where the main stream of the
Zambesi appears to come from the east. Beyond
this to the east and north-east his researches did
not go; but it is known from inquiry, and from
information derived from native travellers, that
several lakes of somewhat large size occupy a portion of the table land, and are connected with the
natural drainage of the continent in that part.

Taking advantage of the existence of an important
fork of the Zambesi, Dr. Livingstone ascended the
Leiba in a canoe, and found that stream tranquil,
and flowing through a rich alluvial bottom, subject
to annual inundation. The land was clothed with
rich vegetation, including lofty trees, and peopled
by buffalo, rhinoceros, and lion, the damp sedgy
pools abounding with alligators, and the country
by no means without human inhabitants. From
the direction taken by this feeder of the Zambesi,
it is evident that the land continues to rise towards
the north parallel to the coast, so that the natural
drainage is towards the interior of the continent.
This is confirmed by the discovery that the source

of the stream so long followed, and the natural
watershed of the district are in the high table land
near the coast, and do not proceed from a lofty
mountain range in the interior. Thus for some
distance before reaching the watershed, which
seems itself to be on level ground, very large
and perfectly flat plains are reached, on which
the water rests during the whole of the rainy season.
The river sources, therefore, are to be found quietly
oozing out of bogs, and not, as in Europe and other
countries, where there is a distinct central axis of
mountain, commencing in some mountain gorge,
rushing down the narrow gullies, and so entering
the lower plains, and constantly increasing till they
reach the sea.   In African rivers a very large part
of the water is evaporated before the sea is reached,
so that streams very important in the interior have
no considerable outlet.

From the pools and stagnant shallow lakes discovered in the upper part of the Leiba, and evidently forming the true and only watershed of the
district, innumerable streams and streamlets run
off to form the great rivers known respectively as
the Zambesi and the Coango or Zaire. These are
two of the principal rivers of South Africa, and
both originate in the same locality and in the same
manner. As, however, the Zambesi has a far longer
course, and drains a much larger country, it is in
all respects the principal river.   These rivers
meander, anastomose, and collect into lakes in the

great central plain of Africa, before they take a direct course to the sea, and a large proportion of the water that falls on the plains is every year carried away by evaporation.

The discovery of the mode in which the African rivers of the south originate leads at once to a determination of the problem as to the form of the interior of the country. Africa resembles an ocean bottom, high about the circumference, but depressed towards the centre. It is not like the other continents, where land rises gradually to a culminating point or ridge in the centre, but is not unlike an important part of Europe—the peninsula of Spain and Portugal—where a similar skirt of high ground rises directly from the sea. Having once attained a moderate elevation, the ground in the Peninsula remains at that level, not sinking in any marked manner towards the interior, and thus differs from Africa in an important sense. It is in connexion with this structure that we must explain the peculiarities of physical geography in Africa which doubtless involve many corresponding changes in all departments of animal and vegetable life.

Dr. Livingstone, after reaching the watershed of the Zambesi, proceeded towards the higher ground near the coast as far as the town of Loanda, crossing numerous rivers, all of which, however, belong to the drainage of the hill country, and empty themselves into the Atlantic. He afterwards returned, and went down stream once more to his former position near Lake Ngami, and thence proceeded to the north-east, after visiting some remarkable waterfalls, the river, having a width of a thousand yards, leaping down a hundred feet over a basaltic ledge, and entering a gorge only about twenty yards wide.

From these falls a hilly tract was travelled over, crossed by numerous streams, some proceeding to the north and others to the south, until, at a distance of nearly 250 miles, the river was again reached. It was no doubt a principal branch proceeding from the west, and soon another branch coming from the south was passed. It appears indeed that the whole country is intersected by streams communicating one with another, and communicating also with the great lake towards the north already alluded to, but which was not visited by Livingstone. His course now lay down stream, and he followed the course of the river, diverging at intervals until he reached its outlet on the east coast of the continent, passing on his way a fine undulating healthy country, swarming with inhabitants, and perfectly accessible.

On the whole, then, the map of Africa, so far at least as physical geography is concerned, has, as we have said, been filled up by Dr. Livingstone and others nearly to the fifth parallel of latitude south of the Equator. Up to that limit the whole country presents the character of a shallow trough, whose edges on both sides are composed of granite and schist, or modern igneous rock, rising to high elevations towards the coast, but rarely attaining the proportions of a mountain chain, and sinking towards the interior, where these rocks are covered either with sedimentary deposits, often of fresh-water origin, or else with recent volcanic products. The drainage is carried on by anastomosing rivers towards large natural pools of very small depth, whence the water makes its way through some fracture of the edge of the basin to the sea, although before this a large part of it has generally been evaporated.

North of the district thus described we are now also provided with accurate records on the eastern side by the exertions of Captain Burton and his associate Captain Speke, who, starting from Zanzibar (lat. 6° S.), advanced westwards across a rich level country for about one hundred and twenty miles, and then reached a hilly district, of which the elevation was about 6000 feet. After this the country further west is for some distance a plateau of 3000 feet, almost a dead level, which descends gently towards the interior as far as the Lake Tanganyika, which was found to be 300 miles long, with an average breadth of between thirty and forty miles. This lake occupies a depression a thousand feet below the plateau, and receives the drainage of a large area. Its distance from the coast is about 500 miles, and its outlet was

not seen. Proceeding northwards alone, Captain
Speke reached the southern extremity of Lake
Nyanza, whose level was found to be higher
than the average of the plateau, and which is
situated directly under the Equator, its southern
extremity being in 2° 30' S., and its length,
in a nearly northern direction, is estimated at
not less than five or six degrees, or nearly 400
miles. This lake is considered by Captain Speke
to be the main source of the White Nile, and to
exclude altogether the possibility that there exists
any group of lofty mountains such as have been
described from the earliest time under the deno-
mination "Mountains of the Moon." In all probability these so-called mountains have received
that designation because the inhabitants of the
district are called by a name which also signifies
" moon," and thus the hills and high ground surrounding the lakes have been also so named. It
is, however, very curious that the direct statements
of some German missionaries—Drs. Krapf and Reb-
mann—and of an Egyptian expedition, who have
been within a hundred miles of the northern end
of the Lake Nyanza, should speak in positive terms
of such a snow-capped range, so that the matter
still remains in doubt. It is possible that this
range may be a very lofty portion of the coast
range, considerably to the north, and perhaps
to the east of the position hitherto assigned
to the Mountains of the Moon. The actual con-

nexion of the northern extremity of the Lake Nyanza with the waters of the White Nile must soon be proved, and this small gap being filled up, there will be nothing left to make out on the whole east coast of Africa, from the sea to the great central depression.

The great problem of Africa may now be said to be solved. There is the great east and west mountain chain of the Atlas running across the continent near the north coast, and corresponding high ground near the east and west coast. This latter elevation forms a boundary wall not generally more than six thousand feet above the sea, extending towards the south-east and south-west parallel to the seaboard, and converging in the high table land of the Cape. Directly south from the Atlas range is the Great Sahara, which is by no means a complete desert, although, being irregularly and poorly supplied with water, it is, on the whole, unfavourable for vegetable and animal life. There are no lofty central mountains of any kind, but in their place a succession of vast plains, south of the equator, which are well watered by an anastomosing system of rivers connected with great sheets of shallow water, varying greatly in dimensions at different seasons.

Almost all the land seems cultivable, and a very large proportion is inhabited by a variety of black races, who are not without an admixture of Arabs, and who seem to possess a limited, but decided, civilization. These people are capable

of commerce, and have abundant material to dis-
pose of. They are not wanting in intelligence;
and if, on the west coast, they have been degraded
by the slave-trade, they seem, where freed from that
scourge, in a state of improvement. So far as Dr.
Livingstone could determine, there was nothing of
that ferocious display of savage nature in the south
that has been found north of the Equator; and the
experience of Captains Burton and Speke on the east
is not dissimilar. If there is less romance than had
been supposed, the difficulties that stand in the way
of progress are fewer, and there is more hope. It
is certainly not extravagant to look forward to the
establishment of commercial relations with the
inhabitants, which will probably lead to other
results equally important; and there is no reason
why the Africa of a century hence may not be well
known and useful to mankind at large.

    There is, in fact, more promise of successful results
in dealing with a people who may be regarded as
in the infancy of progressive civilization than in
forcing commerce on a race who, like the Chinese,
are childishly vain and proud of the small
growth they long ago made, and who are also
earnestly resolved to resist, as long as possible, all
forms of improvement suggested by the "outer
barbarian."

VI.

THE INTERIOR OF AUSTRALIA.

Peculiarities of Australian geography—State of knowledge in 1840—Survey of the coast by Stokes—Eyre's expedition from Adelaide along the coast to the west—Sturt's journey to the Salt Desert, by the Murray and Darling—Leichhardt's first expedition from Moreton Bay to the Gulf of Carpentaria—Discovery of various minerals and of gold—Examination of the south-eastern country—Intervals yet unvisited—Mr. Gregory's expeditions—General conclusion as to the physical geography and geology of Australia.

THE interior of Australia is even less easy to penetrate than the interior of Africa. The great island-continent, compact in form, not crossed, as far as is known, by any lofty mountain chain, is suspected of having, far in its central and hitherto unvisited portion a vast saline desert, cutting off all communication. But notwithstanding the enormous amount of emigration that has taken place since the discovery of gold, and the persistent endeavour of many travellers to determine accurately the condition of this central portion, there is still ample room for adventure.

If we look at Arrowsmith's map of Australia,

dated 1840—just twenty years ago—we shall find
the coast by no means entirely surveyed; the
Colonies in the south-east (New South Wales) and
the south (Adelaide), only to some extent mapped,
and the small belt of land near Swan River Settlement, on the south-west, just indicated. About
that period, however, Captain Stokes, in the *Beagle*
surveying ship, visited the coast, and discovered
several rivers; and in that year also, Mr. Eyre
started from Adelaide with the view of penetrating
as far as possible towards the centre of the country.

He succeeded only in advancing a short distance
beyond the shores of Spencer Gulf to the sandy
flat bounding Lake Torrens, his farthest point north
being in 29° south latitude; but he proceeded along
the coast, connecting Adelaide with Swan River.
Five years later Captain Sturt, following the course
of the Murray River and its tributary the Darling,
for some distance into the interior, proceeded northwards through a hilly country and reached latitude
27°. Finding there a sandy desert, nearly at the
level of the ocean, he also was obliged to return,
and it since appears that the difficulties of further
advance are all but insurmountable, owing to the
total absence of grass and water in that direction,
and the commencement of a saline desert.

Although the single tracks thus made by Captain
Sturt and Mr. Eyre were two hundred miles apart,
and neither of them could be said to advance very
far towards the centre of the country, still the

character of the interior sufficiently agrees. The elevation above the sea was not considerable; the drainage was confined to the rainy season, and the rivers, though occasionally large and full, suddenly, after a few days' drought, became reduced to a series of unconnected puddles. The country seemed almost equally difficult of access at all seasons, the rain converting large tracts into a swamp, and the drought producing almost immediately a complete destruction of the green food, and a drying up of all water-courses. Generally the permanent vegetation is scanty, but there are intervals, or oases, where there is a better growth. Hot dry winds blow from the interior, adding sometimes greatly to the difficulty of travelling, and stony deserts alternate with the low hill ranges.

While Mr. Eyre and Captain Sturt were exploring northwards from Adelaide, Captain Stokes, after surveying the Gulf of Carpentaria in the north, and discovering Albert River, advanced some way into the interior towards the south.

In the year 1844 Dr. Leichhardt started with a small exploring party from Moreton Bay, in lat. 27° S., long. 153° E., with a view, if possible, of crossing the whole of Australia to the settlement of Port Essington, on the north-western extremity, in lat. 11° S., long. 132° E.; and after two years of the most perilous enterprise, during which his whole party was believed to be lost, he succeeded in accomplishing his object, crossing nearly 3000 miles of country, which he found to be in every respect superior to the portion of Australia hitherto visited.

Dr. Leichhardt advanced north-westwards from Brisbane to the east side of the Gulf of Carpentaria, his distance from the coast being generally about 150 miles. He crossed several streams, some of considerable size, and a considerable tract of country, where permanent tree vegetation indicated a better climate than is found in many districts. Still there was a large preponderance of table-land, broken by low mountain peaks, and no promise of cultivable land towards the interior. The streams, like all others hitherto discovered in Australia, were too variable to possess much value for purposes of intercommunication.

Although smaller expeditions were constantly making fresh additions to the knowledge of the interior, and connecting discoveries already made, and whilst the geology of the coast and the country near it was gradually being learnt, there was no important step taken towards the discovery of the interior of Australia for many years. Between 1848 and 1850, indeed, the south-western district was again travelled over, and coal shales discovered; but the country was found for the most part inaccessible, owing as usual to the want of fresh water towards the interior. Numerous small lakes of salt water were intersected, and the country made out with tolerable completeness along a total length of at least 600 miles of coast. Both lead and copper were found, and there were clear proofs of mineral wealth. Shortly afterwards, the discovery of gold turned the attention of all to inquiries which had little to do with direct exploration, and the only parts of the country that were for some time visited were those in which a prospect appeared of finding that precious metal. But for this very reason the whole southern tract of Australia on the eastern side has been examined very carefully, and within a distance of about two hundred miles from the coast may now be said to be occupied, while there is some little known for the next three hundred miles. The extreme range of research is about eight hundred miles from the mouth of the Murray River, in the colony of South Australia, northward to that salt desert which was reached and coasted by Sturt in 1845. A similar desert has now been discovered about the same distance west of Brisbane, and the other towns on the northern part of the colony of New South Wales. All the intermediate country between the coast of the western part of South Australia colony, and the coast of Northern Australia, in latitude 20° S., may now be regarded as known in its principal physical features. It includes a tract of about a million of square miles, having a high coast range, or cordillera, on its eastern side, and detached mountain groups on the south, between which pass the Murray River and its tributaries, the main stream running westward from the Australian Alps, near Melbourne, in Victoria, and the principal tributary—the Darling—coming down from near the north-easterly limit of the districts, and receiving numerous tributaries from the high coast range of New South Wales. East of the Darling and its tributaries lie the whole of the settled districts, while towards the north and west are the poor, imperfectly watered plains tending towards the still poorer and apparently desert country in the interior.

. A vast interval—more than a thousand miles—of unexplored country ranges between the more settled parts of the colony of South Australia (Adelaide) and the Western Australian settlements on the extreme south-west coast.   Nearly a thousand miles of unknown country extends also between the shores skirted by Mr. Earle and those north-easterly shores which are still only imperfectly surveyed. Our maps show not a single river of the smallest importance for 1400 miles between the Murray and the extremity of the south coast (twenty-five degrees of longitude), nor again between Murchison River on the west coast (lat. 28° S., long. 114° E.,) and Victoria River on the north coast (lat. 15° S., long. 130° E.), a distance, in round numbers, of 1600 miles, not counting in either case the irregularities of the coast.   From the latter point Mr. Gregory, ascending Victoria River, and advancing into the interior, reached a watershed of about 1600 feet, and continuing to advance southwards for a distance of about 400 miles, traversed a level, sandy, and rocky country, imperfectly watered by a stream running towards the south, terminating in a dry and sterile tract, covered partly with salt lakes, dry at the time of visit. This desert is about six hundred miles in a direct line, west-north-west from the extreme point reached by Captain Sturt, in his expedition in 1845 from the South.

Round a very large proportion of the west, north-west, and north coast of Australia there is thus a belt of land measuring from sixty to one hundred miles, or more, in width, sufficiently watered and sufficiently fertile to justify settlements being established, while the whole, or almost the whole, of the south-eastern part of the country is already settled and cultivated. The northern or tropical part, though of course subject to the inconveniences of all tropical climates, appears, on the whole, sufficiently healthy to justify occupation. Inside that narrow belt of cultivable land there lies an untrodden area, a great part of it so barren and inhospitable that it seems hardly tenantable by any living animals, except some of those whose peculiar structure marks their adaptability to a country where water is not regularly accessible, but exists only in distant puddles connected by water-courses, dry during the greater part of the year. There would appear to be in this interior a vast desert, not, perhaps, without oases of importance, but at present hardly accessible, and scarcely occupied by constant vegetation. The measurements of this desert, however, if it exist without break, are so large and so utterly incommensurate with the usual proportions of such tracts, as to justify great caution in accepting this hypothesis. The progress of discovery is indeed steadily, though slowly, introducing the settlers to a knowledge of cultivable tracts that may be reached by crossing only moderate distances of desert.

The physical geography of Australia is, therefore, altogether peculiar, and can be compared to that of no known district. Its coast is rich in mineral wealth, and, though naturally covered by a vegetation almost useless to man, is quite capable of producing useful crops. Its interior is a widely-extending low plateau, either not watered at all, or watered by small streams lost in the salt- or fresh-water pools that are probably wide and shallow during part of the year, but dry during the remainder. Here and there a river of small magnitude proceeds from the coast range to the sea, but this is a rare exception. Most of the coast streams are in the highest degree irregular in their supply of water; most of those in the interior are equally irregular, and consist rather of muddy pools occasionally connected than of any continuously running water.

The general geology of this remarkable tract of land is of necessity in close relation to its physical geography. There appears to be a margin of chemically-formed rocks elevated round the whole contour of Australia, much in the way in which the mountains connected with recent volcanic eruptions enclose the Pacific Ocean. Within this margin, near the hill range, and forming also its outer flanks, are old stratified rocks, but at some distance in the interior, with occasional patches of basalt, we find vast tracts apparently of recent sea bottom, the hollows and pools converted by evaporation into salt pans. As a general explanation and illustration of the phenomena hitherto known, it may be said that both Australia and Africa are tracts of sea bottom elevated by a connected series of disturbing movements acting round the margin of a central area; while Europe and Asia owe their peculiar condition to the elevating axis being linear, and running across the middle of the area to be lifted; and America is the result of an axis of elevation parallel and near one side of the area. Africa differs from Australia in having one principal and lofty range on the northern side to which the other two are subordinate.

VII.

STATISTICS OF EARTHQUAKES.

Seismology—Meaning of the term—Mr. Mallett's researches in earthquakes—Matters to be determined about earthquakes—Their nature and different kinds—Loss of human life during earthquakes—Origin of earthquakes—Direction or axis of the earth-wave—Mode and rate of progression of earthquakes—Distribution of earthquake shocks over the earth—Range of large earthquakes—Frequency of occurrence—Excess of earthquakes in the cold season of the year—Theory of earthquake action.

AMONG the hard words that have been introduced into our scientific language within the last few years, may certainly be ranked *Seismology*. It means the science of earthquake phenomena, but we almost doubt whether it is sufficiently adapted to the genius of our language to become popular, even in a scientific sense.

This word, ugly as it is, describes, however, a new science, or rather a field of discussion only just beginning to be pursued with system; and perhaps, after all, like *Palæontology, Odontology, Photography,* and half a score others, may become one of a class of household words to a future generation.

At present it certainly demands explanation; and, by heading the present chapter Earthquakes, instead of *Seismological Phenomena*, we shall perhaps stand excused, even by some of our geological readers, who may possibly not even yet have become so imbued with the classic spirit as applied to science as to change *Secondary* and *Tertiary* into *Mesozoic* and *Kainozoic*, nor to abandon contemptuously an old familiar *wood-cut* for *Lignograph* (!) or *Xylograph*, because it has been found convenient to speak of Lithograph and Zincograph.

Passing by these vexed questions of name—which are, however, often of greater importance than is fancied by those whose ear is tickled by high-sounding words, and who forget the discouragement offered by unfamiliar designations of familiar things—let us return to the subject of our title, Earthquakes; a subject which, thanks to Mr. Mallett in England, and Professor Perrey of Dijon, has been latterly reduced, to some extent, to a science of observation and calculation. Mr. Mallett, some years ago, commenced collecting together all the accounts of earthquake phenomena he could find, and, by cataloguing them systematically, hoped to obtain some useful generalization. His results were offered to the British Association, and his lists are published in their Reports for the years 1850, 1851, 1854, and 1858 respectively, the records reaching only to the end of 1842, as since then the published catalogues of M. Perrey give all that is required up to 1850. Including the two catalogues, we have before us accounts, more or less complete, of between 6000 and 7000 separate earthquakes, commencing 1606 B.C., and therefore ranging over 3456 years.

It will readily be understood that, although even within the last ten years of observation the accounts are far from complete, inasmuch as there are no scientific, or, in any sense, available observations over enormous tracts of land, and scarcely any means of ascertaining shocks that commence far out at sea, still the proportion that the number recorded in the last half century bears to the real number that have taken place is far nearer the truth than that of former times. The later returns, although incomplete, are tolerably to be depended on for Europe and the adjacent islands, and it may also be assumed that the matters observed, and the time of occurrence of shocks, have been more carefully noted.

There are many things to be considered with regard to earthquakes. In the first place, What are they? Secondly, Where do they come from, and what direction do they take? Thirdly, How and at what rate are they propagated through the earth? Fourthly, What parts of the earth are most subject to them, and how do they stand related to volcanos? Fifthly, How often do they occur?

Sixthly, Are they at all periodical in number or intensity? and if so, at what times do they seem most abundant? And lastly, What kind of explanation can be given of their origin?

To this long list of queries we will endeavour to give a few replies.

And, first, What are they? This is a question apparently easy to answer in a popular sense, but yet not without its difficulties. We may reply, that an earthquake at any spot is an undulating or wave-like up-and-down motion of the earth, very nearly resembling the corresponding motion in water when a part of the water is suddenly raised up or sunk down, and the wave or motion thus produced is propagated towards the margin of a vessel holding the water, without the water itself being moved laterally at all. A slight experiment made in a trough, by partially lifting with a string a flat stone at the bottom, and putting small pieces of cork above and around, will very clearly illustrate this kind of motion, and the fact that the wave or disturbance travels; but the particles of water only move up and down, and are not moved horizontally. So in the earth, if from any cause an elastic surface is lifted by a force from below, a similar wave is produced, and the surface rises and falls over a gradually increasing circle, the magnitude of the wave constantly diminishing as the diameter of the area of extension becomes larger.

The oscillations alluded to appear to travel in

some one direction, which must be regarded at any place as the true direction of the earthquake there. All kinds of results producible by violence may accompany earthquakes; for it is evident that cracks may be formed if the rocks are inelastic; surface rocks may be broken, or may be thrown off their balance; houses, and churches, and other buildings, may be thrown down; rivers may be swallowed up in open cracks, or may change their beds; and permanent alteration, elevation, or depression of the earth's surface may result. Springs of water or jets of gas may rise out of the crevices formed; and, if the action takes place at the bottom of the sea, water-waves as well as the land-wave may be produced. The waters of the ocean may then rush in on the land far above their usual level, or may retire far below the point usually recognised as low-water mark. Noises caused by the movements of rocks beneath the surface may also be produced, and men and animals may be affected either by the actual motion produced, or by peculiar atmospheric or other conditions incident to the disturbance going on.

Such are some of the principal phenomena of earthquakes, but all of them are not noted in every case. There are, indeed, examples of every kind, some being vibrations so small that they are not readily distinguishable from the rumbling of a heavy waggon over stones, while others are, without exception, the most awful and destructive events recorded in history. There is sometimes a single vibration, terminating at once; but oftener the earthquake consists of a series of undulations (with intervals of repose), lasting for hours, days, and even weeks. There is one curious fact regarding their effect on human beings; for, instead of those who are subject to them, and who live in districts where they are common, becoming indifferent and careless about them, the exact contrary is the case. To strangers they are matters of curiosity and interest, but to the inhabitants objects of unmitigated horror.

There are different kinds of earthquakes; first, those which are mere undulations, heaving the ground at any one place upwards and downwards, and producing the same result at many places along a certain course over the earth; secondly, those which are like the explosion of a mine, and consist of a sudden upheaval without undulation; and, thirdly, those which are complicated of the advancing wave and the direct upheaval, and result in a peculiar kind of twisting motion, not unlike that of a ship in a cross sea, apparently much more destructive to human constructions and human life than either of the two simpler kinds.

The calculations that have been made as to the destructive character of these disturbances fully warrant the remark made above, that they are of all terrestrial events the most fearful.

Thus it is on record that upwards of 60,000 persons perished in the great earthquake of Lisbon

in November, 1755 ; 10,000 in another in Morocco ;
40,000 in Calabria ; 50,000 in Syria on one occasion, and probably 120,000 in the same country, in
the time of Tiberius, A.D. 19.   In the year 526,
250,000 persons are said to have perished at
Antioch ; and seventy-six years afterwards, a second
earthquake destroyed 60,000.   At Messina, in
1692, 74,000 persons are said to have perished, and
in Quito, in 1797, 40,000, although the population
of the province was then small.

So great is the number of victims believed to
have perished during earthquakes on a large scale,
that if we take the average of important earthquakes
to be one in every eight months, and assume that
about one in four seriously affects human life—and
we shall see hereafter that this average is fully
justified by observation—it will appear that several
millions of human beings have been suddenly
swallowed up in this way within the last four
thousand years.   There is reason to believe that
very large quantities of animal remains and vegetable
matter have also been entombed by the earthquakes
that have occurred from time to time, even when the
number of human lives lost has been much smaller
than above recorded.*

Where, then, it may be next asked, do these
earth shakings proceed from, and what direction
do they seem to take ?   The illustration given of the

---

* See Report of Meeting of British Association, 1851, p. 63.

nature of the earth-wave, and its resemblance to a certain kind of water-wave, will already have suggested an answer to this question.

They do not necessarily come from the direction they have at any one part of their progress, for the wave may have been broken, and may appear to come from another centre than that from which it really does proceed. It is only by multiplied observations, and comparisons of the same earthquake in different places, that any satisfactory result can be looked for in this inquiry, and such observations hitherto have been very few. Still, with the clue now possessed, the source of disturbing action has been approximated in some instances, and has been found sometimes much less distant, but sometimes much more so, than had been imagined from the nature of the shock. It is not every great shock that is close at hand, nor is every lesser shock the dying out of a larger wave originating at great distance.

As a general rule, though one not without many exceptions, it would seem that a considerable area is shaken by each disturbance; but that this varies from what may be called a great earthquake, where the area may be taken at an average as a circle of six hundred miles radius, to a small one, where the radius is not more than seventy miles. Extreme cases much beyond these limits are rare. Small earthquakes, on the other hand, are very common.

The direction of earthquakes seems to have some

reference to neighbouring mountain chains and
coast lines, and it seems clear that the greatest
intensity of earthquake action is everywhere, as
might be expected, identical with the lines of
greatest activity of volcanic action; but having
said this, there yet remain a vast number of smaller
earthquakes that seem to have nothing to do with
volcanos, so far as we know of their existence.
Mr. Mallett, indeed, gives as his opinion that " an
earthquake in a non-volcanic region may be viewed
as an uncompleted effort to establish a volcano,"
and this may be regarded as one of the results of
investigation.

Notwithstanding what has been said as to the
expansion of the wave in concentric rings, it does
not appear that earthquake shocks are felt so far
in some one direction (which may be called breadth),
as in some other which we may therefore regard as
length. Thus the space disturbed is rather oval
than circular, and the longer axis of the oval is
generally identical or parallel with the lines of
elevation which enclose the great oceanic basins of
the earth's surface. This seems to be the case
whether such basins are marked out by actual land
above the present sea level, or only by submarine
elevations or shoals, rising from the ocean floor, just
as mountain chains rise from the general level of
the land of the earth.

. It would seem that all principal or first-class earthquakes occur along such lines of elevation, and that as

the origin of the disturbance departs from them, the disturbance itself becomes less important in its phenomena and results. On the other hand, it appears that the spaces enclosed within these lines of elevation, whether above or under water, are for the most part free from disturbance, and are never the seat of destructive shocks.

We come next to the inquiry as to how and at what rate an earthquake is propagated through the earth ? But as under the last heading the nature of its wave-like progress was alluded to, it will only require here to be somewhat further illustrated.

In a general sense, the earth's superficial crust may be regarded as an elastic medium, capable of having waves propagated through it; but in fact the materials, even within very narrow limits of space, are so varied, their condition with regard to hardness, density, elasticity, and other important points is so different—some are so loose and open, others so tough and close—some consist of such solid masses of the same nature throughout, while others are made up of beds, each of such different material, and of such irregular thickness, that no approach to systematic structure can ever be obtained. Thus an earth-wave, which is produced and travels at a certain rate in some one rock, must as it advances traverse an infinite variety of material. It soon therefore becomes broken up into a multitude of smaller waves, each one of which not only pursues its own course independently, but is itself soon

I

subject to be divided, and so on till the whole effect
has died away.

On a very large scale, as in the great earthquake
of Lisbon, and sometimes on a smaller scale, where
several observations are recorded pretty accurately
of the same disturbance, it has been possible to
calculate the rate of motion of the earth-wave ; but
this rate is found to vary so much, that no average
can be given, except for each individual case. Thus,
in the case alluded to at Lisbon, in 1755, the
general average from seventeen recorded observations
at distances varying from thirty to twelve hundred
miles, the rate was sixteen miles per minute, and this
agreed pretty well with each particular case of the
seventeen. In the second great earthquake of
Lisbon, in 1761, felt equally widely, but not, it would
seem, through a larger area, the rate was thirty-four
miles per minute, the number of observations being
seven. And in an earthquake in California in 1857,
noticed for a distance of two hundred and fifty
miles from San Francisco, but carefully observed in
five places, the rate was only between six and seven
miles per minute. It is evident, therefore, that the
circumstances in each case must differ very widely.

M. Perrey has endeavoured, from the materials
at his command, to determine a sort of mean
direction of the various earthquakes occurring
within certain geographical limits, especially in
Europe. Some of these mean directions seem to
correspond pretty well with those of the river valleys

of the districts to which they refer, but long series of accurate observations are still wanted before any important generalisation can safely be made.

The parts of the earth subject to earthquakes are very much larger and more widely spread over the surface than is generally imagined. We subjoin a list of countries, topographically arranged, with a statement in each case of the total number of earthquakes observed in them. The observations refer first, to the whole time from the commencement of the record to the end of the last century, and secondly, to the first half of the present century :—

	Number of Recorded Earthquakes		
	To end of 18th cent.	Half 19th cent.	Total.
In the European hemisphere :—			
The Scandinavian peninsula and Iceland	139	113	252
The British Islands	124	110	234
The Iberian peninsula	135	85	220
France, Belgium, and Holland	491*	211	702
The basin of the Rhone	111	81	191
The basin of the Rhine and Switzerland	384	173	557
The basin of the Danube	173	145	318
The Italian peninsula	884	478	1362
Algeria and Northern Africa	—	—	63
The Turco-Hellenic peninsula with Syria	373	197	570
The basin of the Atlantic			140
In the American hemisphere :—			
Canada and the United States	98	51	149
Mexico and Central America	37	30	67
The Antilles	112	195	307
Chili and La Plata	24	170	194

* Of these 237 were in the eighteenth century.

Scandinavia is so intimately related to Iceland, the land of volcanic action, that no one can be surprised at its presenting earthquake phenomena; but although a large number of shocks are recorded (on an average, one in six months during the present century), and the land itself is undergoing a slow but steady process of elevation, the destructive power is not great. The commotions of Norway are always connected with volcanic disturbances in Iceland, and thus relief is immediately obtained without mischief.

The general direction of earthquake action in the British Islands appears to be from south to north, veering more or less to the east or west. This is believed by Mr. Mallett to be in the line of the focus of the Lisbon earthquake and the Canary Islands. Another investigator into these phenomena (Mr. David Milne) believes that the central point of disturbance is immediately beneath our island, and at no great distance, but this idea is combated by Mr. Mallett, and does not seem probable. Besides the movements that are properly recognised as earthquakes, a vast number of little shocks have been almost continuously felt for a long period in and about Comrie, in Scotland, but these are probably no more than tremors, caused by the fracturing of rocks below, when processes of elevation or depression are going on. Most of these quakes have had a direction from west to east.

Spain and Portugal are remarkable as being the

seats of extraordinarily intense earthquake action. The area affected includes a part of the bed of the Atlantic, and reaches as far as the Azores, and the focus or centre of action seems to be beneath the sea, between Lisbon and the Azores. The great Lisbon earthquakes, about a century ago, were among the most remarkable and wide-spreading of any recorded in Europe. The general direction is, S.E. by E. to N.W. by W.; but in the Pyrenees it alters, and approximates to the bearing of that mountain chain, becoming nearly E. to W.

Belgium and Holland, and France (excluding the basin of the Rhone), are very subject to earthquake action, the direction being for the most part along the lines of the principal river valleys. At St. Jean de Maurienni, in Savoy, in the year 1839, from January 27th till June 16th, there were noticed as many as forty-nine shocks, of which nine were rather severe, but the rest moderate and slight, or scarcely perceptible, like those occurring at Comrie, and attributed to the same cause; and many of the disturbances recorded were not widely distributed, although others have been quite unmistakeable, and have done mischief.

In the basins of the Rhine, Rhone, and Danube, as in other similar districts, the direction of the earthquakes hitherto recorded was more or less that of the river valleys to which they were limited; but the local exceptions are numerous and important. Few of the earthquakes recorded are of very great intensity, though buildings have been destroyed and subterranean noises heard.

The Italian peninsula and the adjacent islands are as remarkable for their earthquakes as for the extraordinary intensity of volcanic action in Etna and Vesuvius. Some of the most fearful and mischievous disturbances on record have originated here, as proved by the absolute verticality of the shock at some one spot. The advance of the wave, where so immediately over its origin, is not clearly defined enough to give any general direction. The adjacent country of Algeria and other parts of Northern Africa, are, as yet, almost without records of earthquakes, but such events are both frequent and serious now, and have certainly been so in past ages. The central point of disturbance is quite unknown. Even in the last-named European region, that of the Levant, although fearful paroxysms have there occurred at no distant period, we have no facts that admit of being taken for generalization. There appear here to be several centres of action, besides one in South Arabia, one in the Caucasus, and one in Greece.

Within the Atlantic Ocean, no less than five great and probably connected centres of volcanic action exist—Iceland, the Azores, the Canaries, the Cape de Verds, the West Indian Islands—besides many other points (Ascension, St. Helena, St. Paul's, &c.) at which extinct volcanic phenomena are visible. A remarkable submarine volcanic tract has been recently added to these by M. Daussy, forming a belt about seventy miles from the Equator on the south side, between the 20th and 25th meridians of west longitude. This tract is nearly midway between Cape Palmas, on the west coast of Africa, and Cape San Roque, on the east coast of South America. Besides the principal points determined, are others about one hundred miles to the south, and others again both to the west and east. There are not less than fourteen well recorded observations of earthquake or volcanic disturbance in this area, occurring between the years 1747 and 1836, besides some others not quoted by Mr. Mallett. On some of these occasions temporary shoals were formed, and direct evidence given of eruption; but in others the proofs were limited to indications of submarine earthquakes.

In North America there seem to be frequent shocks, few of them passing further north than Canada, and ranging generally from south to north. In Central America are decided centres of earthquake action, but both here and in the West Indian Islands the north and south direction seems to prevail. This is also the case on the west coast of North America, so remarkable for its volcanos and for its numerous volcanic eruptions from lofty mountains.

It will be evident from the above, that although by no means confined to volcanic districts, earthquakes are very closely related to volcanos. There are, in fact, as already observed, different ways in which subterranean disturbing force shows itself. Thus in a district where the rocks are already fractured, and an easy way is found to the surface, the earthquake does not occur, but an eruption takes place. In a spot where no such facilities exist, the ground is shaken more or less, according to the force exerted, the depth of the centre of action, and the elasticity of the rocks.

Lest the reader, hearing thus of earthquakes and rumours of earthquakes in those parts of the world he is accustomed to regard as safe from such danger, should feel uneasy and begin to tremble in anticipation, it may be well to occupy a few lines with a list of those districts where serious earthquake action is not only possible, but in some measure, judging from experience, may be regarded as probable. These are the lines and areas of greatest intensity of earthquake, in contradistinction to those of smaller intensity, in which latter England, and the rest of Northern Europe, have long been included.

Intense earthquake action has been confined in Europe, in modern times, to the island of Iceland, the Atlantic coast of the Iberian peninsula, and the South of Italy. In Asia it includes the shores of Asia Minor, the north-western and north-eastern extremities of the Indian peninsula, the coast of Siam, the whole line of islands of the Indian Archipelago, and the islands ranging northwards, (including Japan,) parallel to the coast of China. In America, the whole tract of country on the west coast, from the south of California to the islands south of Cape Horn, together with the West Indian Islands, and the north coast of South America. A large part of the bed of the Atlantic, and various parts of the Pacific, are also subject to similar disturbances. To a considerable, but inferior extent, almost the whole of Central Asia, and India, both north and south of the Himalayan chain, are dangerous, and the same may be said of some portions of the coast of Africa. It would seem that generally the borders of the Pacific Ocean and of the Caribbean Sea may be looked on as the districts most exposed, with the exception of the chain of islands of the Indian Archipelago.

The other districts subject to earthquake action are not, so far as experience informs us, liable to the more severe shocks. It is, however, quite impossible to calculate the course of action of those subterranean movements and forces that produce a great mechanical result, and we dare not, therefore, limit too closely these areas. It must also be remembered that the evidence ranges only over a period very brief compared even with the duration of the last geological conditions of the surface.

We come next to consider how often earthquakes occur. This must of course depend much on the district, and the reader will be able to calculate the average for himself for the European districts from M. Perrey's table, already given. But we are enabled to generalize better by the aid of Mr. Mallett's catalogue, which is more extensive, and reaches to 1850. We thus find that, while only 787 earthquakes are recorded from the earliest period to the close of the fifteenth century, and only 2804 in the three next succeeding centuries, there are no less than 3240 of which more or less detail has been given during the half century that concluded in 1850. It cannot be supposed that all or nearly all that have occurred were recorded, for there are many wide tracts still unobserved. We know, therefore, that, at any rate, *more* than sixty earthquakes have taken place on an average in each year; and of this number, about three in two years, or one every eight months, has been what may be called a great earthquake—by which is meant one in which whole cities and towns, or large portions of them, have been reduced to rubbish, and many lives lost.

Taking, next, the more detailed observations of M. Perrey, we find the mean annual number of earthquakes recorded, with dates, in Europe and the adjacent parts, between the years 1833 and 1842 inclusive, to be nearly at the rate of thirty-three per annum. It is considered that one-fifth more may have occurred, and have escaped knowledge; the mean annual number is thus raised to forty, or about one every nine days. One cannot help being struck to find so serious a matter as an earthquake systematically reduced to the level of a nine days' wonder.

That earthquakes are periodical in some sense—that they recur at intervals in some measure comparable—and that the number occurring within a fixed period of time is the same, or nearly the same—had been thought probable by geologists, but, till the earthquake catalogues of M. Perrey and Mr. Mallett were published, it could never be proved. Observations, however, having now been arranged and tabulated, we find not only these but other curious relations existing between earthquakes and known periodical phenomena, especially with regard to the age of the moon, the month and season of the year when earthquakes most frequently happen, and the length of the intervals between convulsions of great violence.

With regard to the phases of the moon's motions, M. Perrey found that in the four years, 1844 to 1847 inclusive, the number of earthquakes near new and full moon exceeded the number at the quarters very nearly in the proportion of six to five. In a number of exceedingly elaborate calculations, M. Perrey endeavoured to show that, however the figures were handled, they always presented the same general conclusion.; but there are not as yet sufficient facts to justify more than an allusion to this curious speculation. It does, however, appear to be an inevitable deduction from the evidence, not only that earthquakes occur more frequently at the periods of new and full moon, but that their frequency increases at the time when the moon is nearest the earth, and diminishes when it is most distant; and, moreover, that earthquake shocks are more frequent when the moon is near the meridian than when she is 90° away from it.

Tabulating, next, the various shocks in the months in which they respectively occurred, (regarding each group or succession of small shocks connected together as one earthquake,) and afterwards collecting the months into seasons, we find the following to represent the state of the case when all the observations made in the northern hemisphere are arranged so as to show the numbers during the cold and warm seasons respectively. It will be understood that this table includes the whole number of earthquakes recorded, whenever the record gives sufficiently accurate data :—

April . . . 489		
May 438		
June 428	Warm Months . .	2721
July 415		
August . . . 488		
September . . 463		
October . . . 516		
November . . 473		
December . . 500	Cold Months . .	3158
January . . 627		
February . . 539		
March . . . 503		

Such a calculation might be the result of grouping together a number of cases which, if taken fairly, each in its relation to its own district, might show a different result. We will next, therefore, take M. Perrey's table of the European earthquakes, in his list recorded between A.D. 306 and 1843. Without particularizing the months—which, however, follow nearly, though not quite, in the same order—and taking separately into account the earthquakes of the present century as being the most trustworthy, we have the following result for Europe :—

	To end of 18th cent.	During 19th cent.	Total.
Warm months	394	463	857
Cold months	525	638	1153
	919	1091	2020

Showing that in the European list the excess of shocks in the cold months is even larger in proportion, amounting to more than one-seventh of the whole number. In other words, for every three earthquakes that are felt in Europe in warm weather, four are felt in cold. This very remarkable result is fully borne out, though not always precisely in the same proportion, by all the separate lists tabulated for the various districts already mentioned in a former page. Thus, out of 217 in the British islands, 94 were in warm and 123 in cold months. In the Iberian peninsula, out of 201, the numbers are 87 and 114 respectively; in the Italian,

out of 993, they are 455 and 538; and in the
French district, out of 667, we have 272 warm
and 395 cold. In the Levant, indeed, the total
number recorded being 436, there appear 222 in
the warm months against only 214 in the cool;
but, if we take the earthquakes of the present
century, which amount to 196 (nearly half the
whole number recorded), we find the same excess
as in the other districts—the cold months giving
103, and the warm only 93. In the doubt that
exists as to the real value of the tables before the
year 1800, the latter must be regarded as the
nearest approach to an average.

In the Southern hemisphere, where the climates
are, of course, reversed, we find a general indication to the same effect, although the number of
observations as yet is too small to have much
value.

Another curious result appears from these tables.
Regarding the periods near the equinoxes (March
and April in spring, and September and October
in autumn), and the solstices (June and July in
summer, and December and January in winter), as
" critical periods" of the year, we find that, during
the first forty-three years of the present century, the
recorded earthquakes are as follows (the average
number of shocks for two months being 152) :—

Spring equinox	151
Summer solstice	129
Autumn equinox	164
Winter solstice	177

There are some other generalizations, obtained from an examination of the earthquake tables which are worthy of remark.

In the first place, earthquakes are paroxysmal phenomena, following no absolute law, but yet regulated, and showing on a large scale a system of averages and compensation. Small earthquakes occur often at small intervals; the average of those not absolutely unimportant in their effects being, it would seem, from five or ten years, during which interval there is comparative repose.

The shorter intervals seem to be in connexion with periods of fewer earthquakes, and usually, but not always, those of least intensity.

Great earthquakes seem to have occurred for some centuries past at intervals of about a hundred years, and groups of several important convulsions at intervals of fifty years. Thus, within the last four hundred years, we find that the middle and latter part of the sixteenth century was marked by great and numerous earthquakes in China, Europe, and the Atlantic, many of them very severe. In the middle of the seventeenth century, there were great and disastrous shocks in the Mediterranean basin; and towards the latter end of it occurred the great Jamaica earthquake, besides many others of importance. Towards the middle of the eighteenth century was the great Lisbon earthquake, and subsequently the great one in Calabria. Hitherto, during the present century, there have been none of very extreme intensity; but they may perhaps be looked for before long. There thus appears to have been an interval of about a century between each of the very greatest paroxysms; and a like period may be traced between those of next importance in each century, following the former at an interval of from thirty to forty years. It also appears that, near the time of the great paroxysms, a number of smaller, but still important ones have been crowded into four or five years; while, near those of second importance, a number also large is thickly spread over ten or twelve years. As the record of the greatest disturbances is of course more likely to be found in history than that of smaller ones, it seems further worthy of remark that the first, fifth, ninth, twelfth, and eighteenth centuries of the Christian era seem to have been those when the destructive force of earthquakes has exercised the largest influence over the human race in civilized countries; while the first and second A.C., and the third, seventh, tenth, and fourteenth B.C., of our era, were times of comparative repose.

The last inquiry in this matter is concerning the theory of earthquake action. It is also the most difficult and the least developed, and will not detain us long.

It may be stated in a general way that earthquakes are apparently the result of impulsive forces acting at an unknown and variable depth below the surface of the earth, and may result in volcanic eruptions if in any case the force is sufficiently great to fracture the rocks above. Assuming, as is generally done,* that the interior of the earth is in a liquid or melted state from heat, this interior, like the waters on the surface, may be supposed to bulge out under the influence of the moon's attraction as it revolves round the earth, thus producing a tide in the liquid matter within. It is on this possibility that the observation of earthquakes in reference to the moon's position becomes interesting and important. M. Perrey imagined four methods of calculation applicable to the materials he accumulated, and although we cannot here give even an outline of the methods, as they are too abstruse for a popular summary, we may repeat that the conclusion arrived at by him, and supported by the investigations of another observer, shows that earthquakes are most frequent when under this view they ought to be, and least frequent when the calculation would make them so. That any direct connexion is thus proved is however more than the figures warrant us in asserting; although they may be said to support the conclusion that changes in terrestrial temperature, or in

---

* This assumption, although commonly made by a large class of geologists, is not to be taken as in any sense proved by observation or experiment. It is here taken as a theory supposed to be strengthened by the results of inquiry into the phenomena of earthquakes.

K

the circulation of electric, thermic, or magnetic currents, converting force into heat, may produce the phenomena recorded, and that the interior mass of the earth may be affected by atmospheric or terrestrial tides.

No doubt can exist that both sun and moon, both directly and indirectly, do largely influence our planet, and that even such phenomena as the periodicity of solar spots require to be taken into consideration if we would bring into one group all the facts that bear on any such phenomena as we have been considering. How, when, and in what time the changing influences may act, it is not yet possible to say ; but the progress of science cannot fail to throw light on these as on other difficult matters of cosmical investigation.

VIII.

ORIGIN OF VOLCANOS.

Opposite views entertained as to the origin of volcanos—History of the formation of Jorullo — Humboldt's views — Scrope and Lyell's views—Objections to the old theory—Mode of formation of a volcano according to this view—Absence of necessity of upheaval—Frequent destruction of the crater and mountain top by an eruption—Renovation of the mountain—General absence of dislocation on a large scale in volcanic districts— Shortness of duration of a volcanic district in comparison with other geological events.

A GREAT controversy has existed amongst geologists for the last thirty-five years as to the cause of the peculiar form of volcanic mountains, the origin of volcanic cones and craters, and the nature of the force by which such phenomena have been produced. Great and influential names have been numbered among the supporters of the two theories put forth, and even at the present day the differences of opinion are very marked. On the one side, which may be called the rationalistic, are ranged Scrope and Lyell among our own countrymen, Constant Prévost in France, and most of the early describers of volcanic phenomena abroad, as Saussure, Dolo-

possibility of accumulating lava poured out in a
fluid state in beds pitching at a high angle, and
ranged round the cone, as they are often found,
especially in Etna; and third, the ordinary condition of the crater in the case of every volcano—
this condition being said not to admit of full explanation by any other theory than that of elevation.   It should be stated that all the lava currents
of which the solid part of the cone of a volcano is
formed are assumed, according to this view, to have
been deposited horizontally under the sea at some
depth, before undergoing the process of elevation,
and therefore before the commencement of the
existence of the volcano as a mountain vomiting
fire.   The French geologists (usually inclined to
paroxysmal and convulsive views) assume the almost
instantaneous production even of the large volcanic
mountains, and even Von Buch, generally cautious
enough, does not hesitate to advocate similar
views.

In a Memoir recently published in the Journal
of the Geological Society,* Mr. Scrope combats
some of these views; and another Memoir, by Sir
C. Lyell, in the Transactions of the Royal Society,†
contains a review of the whole subject, together
with additional evidence in support of the view he
had taken in 1828, and which was advocated in the

---

* Quarterly Journal of Geological Society, vol. xv. pp.
505—549.
† Vol. cxlviii. p. 703.

first volume of the " Principles of Geology," published in 1830. It will be interesting, and may be
useful, to compare the arguments on both sides,
though without copying the somewhat excited tone
of Mr. Scrope.

. The facts recorded concerning Jorullo, which, it
must be remembered, was not seen by Humboldt
till twenty years after its formation, and of which
no written account was prepared at the time,
although the events were subsequently narrated to
Humboldt by eye-witnesses, do not seem really in-
consistent with the views of Sir C. Lyell and
Mr. Scrope. There appears to be there a wide
tract of basaltic lava, out of which one large and
several smaller cones of eruption have risen, these
cones being formed of scoriæ, ashes, and other
fragmentary matters erupted, while the large surface of lava is connected with the larger cone by a
bulky promontory of coarse-grained lava. The
greatest height of the cone is twelve hundred feet
from the plain, and that of the lava at the foot
four hundred and seventy feet, which is not more
than the known thickness of some of the lava
streams of Iceland. The blisterlike appearance
described is perfectly compatible with the lava
having poured out in a pasty state from the earth
over the flat plain, naturally accumulating near the
point of eruption, but also running off on all sides,
and becoming gradually thinner.

Humboldt also describes and was much struck

by a number of very small smoking cones (*hornitos*, or little ovens, as they are called by the Mexicans), under ten feet high, covered with or composed of decomposed basalt. But these seem to have become obliterated after some years, and there is no reason for supposing that they were ever more than heaps of volcanic ashes mixed with rain water, made to boil by the hot lava below. Their structure and the steam that issued from them, are thus fairly accounted for.

The case of Jorullo would seem to be the only one on which the advocates of the elevation theory rest as excluding all idea of the older and more simple view of the formation of the cone by the constant addition of erupted matter. If, therefore, a fairly probable explanation of the phenomena can be given, on this latter hypothesis, the whole question must be regarded as open to controversy, and ought to be decided by careful investigation of volcanic phenomena generally. It is pretty clear that there is really nothing in Humboldt's own account which absolutely settles the question, even when we admit, to its full extent, the great value that must ever attach to the opinions of a man who was at once the most accomplished naturalist, the most intelligent and indefatigable traveller, the most honest and conscientious observer, and the most complete philosopher of his day. The peculiar blisterlike appearance he speaks of as evidencing the upheaval might just as well have been caused by the welling out of the liquid or pasty lava from the earth in vast quantities on a level plain by upheaval from below; the small cones were probably heaps of ejected scoriæ, hardened by rain, and heated by the uncooled lava beneath, while the principal cones do not seem to have been referred, even by Humboldt himself, to any cause but the erupted matter (or at any rate more than this cannot be proved), so that the whole account, stripped of the theory, resolves itself into a description of facts admitting of more than one explanation, and a series of inferences which, notwithstanding the high reputation of their author, must not be taken at more than their proper value. It so happens that since the date of Humboldt's observations a somewhat parallel case has been described as occurring in Kamtschatka, which is thus described :—

"In July, 1827, a vast stream of lava descended from the rim of the great crater of Awatscha—a volcano in Kamtschatka—and after pouring down the outer flank of the cone, spread out widely at its base in a high platform, which is covered with a thick bed of ashes. Small hillocks, ten or twelve feet high, rise out of this bed, and emit steam and gas."

· Let us then proceed to consider such mechanical and palæontological objections (these being in fact all that ought to be looked at) as have been at any time brought forward against the views of the

earlier geologists, concerning the origin of volcanic
cones by simple eruption. If it can be shown that
the position of the lava on the flanks of these cones
is incompatible with what is known of the material
itself, or if the appearance of the crater requires
another explanation, or if, finally, there are proofs
found on the slopes of volcanic cones that the sea
has formerly washed them and deposited marine
remains, we may proceed to investigate the
paroxysmal theory; but if it should appear that
lava certainly can be, and is, deposited in a pasty
state at an angle of from 10° to 30°, or even in extreme cases of 35°, without running down and
moving away from the cone of eruption, if also
craters can be shown to be, in some cases at least,
the direct result of volcanic eruptions in air, and
if there are no proofs of the sea having washed
the flanks of the cone, it will be far more reasonable to accept a simple explanation, agreeable to
common sense, than seek for one far more complex,
and not justified by what we know otherwise of
volcanic phenomena. It is always something
gained when we are able thus to relieve science
from assumptions contrary to experience and actual
knowledge, and bring back within the group of
ordinary events, explicable by the results of experience, any phenomena that have seemed exceptional.

At the same time, the real difficulties of the case
must not be concealed, and may be briefly summed

up as follows :—First, volcanos are often complete and very lofty mountains; secondly, they are connected with tracts of igneous rock, spread over thousands and tens of thousands of square miles of the earth's surface; thirdly, they terminate upwards with shifting cup or saucerlike hollows, sometimes enlarging, sometimes closing, sometimes filled up, and sometimes hollowed out—these being the characteristic features of true volcanic craters; fourthly, there are accounts of fossil shells discovered, either on the slopes, or buried among the materials which form the slopes, of some volcanic cones; and lastly, the cones themselves, as has been previously mentioned, are variously built up of lava and ashes in all possible varieties of condition, usually inclined to the horizon at angles varying from 5° to 30°.

The first question is the possibility of such eruptions as shall form the larger class of known volcanic cones. It is not, of course, imagined that these were produced at one eruption, for every one knows the periodical character of volcanic phenomena, and the repetition of the same process of throwing out ashes, scoriæ, and melted stone at frequent intervals. With reference to this, there is nothing better than to show what has occurred under the personal observation of respectable witnesses, and we have one very good illustration in the account of the formation of the Monte Nuovo, one of the subsidiary cones of Vesuvius, in the year

1538. An eye-witness thus describes the phenomena :—" Stones and ashes were thrown up with a
noise like the discharge of great artillery, in quantities
which seemed as if they would cover the whole
earth; and in four days their fall had formed a
mountain in the valley between Monte Barbaro
and the Lake Averno of not less than three miles
in circumference, and almost as high as Monte
Barbaro itself—a thing incredible to those who
have not seen it, that in so short a time so considerable a mountain should have been formed."
" Some of the stones were larger than an ox.
They were thrown up, the larger ones about a cross-
bow's shot in height from the opening, and then
fell down, some on the edge of the mouth, some
back into it. The mud ejected (ashes mixed with
water) was at first very liquid, then less so, and in
such quantities that, with the help of the afore-
mentioned stones, a mountain was raised a thousand paces high on the third day. I went to the
top of it and looked down into its mouth, in the
middle of the bottom of which the stones that had
fallen there were boiling up just as in a great
caldron of water that boils on the fire."

But this outpouring of ashes is not the only
process adopted by nature. Alternately with such
eruptions, or mixed with them, there are poured
forth vast floods of lava in a fluid or pasty state.
These generally find vent at first near the bottom
of the cone, and then, as that accumulates, they

burst through its sides, and occurring at intervals, and generally from different parts of the cone, because the resistance offered by ashes is less than that by old and hardened lavas, the result is a mixed construction of the cone, which consists partly of the lava and partly of the ashes through which it has passed, or on which it has been deposited. The quantity of the lava will be understood by the accounts given of a single eruption in Iceland, that took place in 1783, when a current of lava not only filled, but overflowed, a river gorge which was, in many places, 400 to 600 feet deep, and 200 feet broad, and then leaving the hills, it filled up a large lake, flowing on constantly, pouring over a lofty cataract, and filling up in a few days an enormous cavity which the waters had been hollowing out for ages. Nor was this all. A short time after the eastern part of the island had been desolated by the torrent described, another river was filled up and the country overflowed with lava in another direction to the extent of about four miles. Thus two streams of lava were poured forth, one fifty miles long and twelve to fifteen wide, and the other forty miles by seven, the depth varying from ten feet in the open plains to six hundred feet in the gorge. At least 40,000 millions of tons of melted matter must have been poured forth during this single eruption—a quantity sufficient to cover four millions of acres a yard deep.

As a further example of the quantity of matter

that issues from a volcano on the occasion of an eruption of importance, but not of the first magnitude, we may mention briefly the particulars of the Etna eruption of 1852-3, beginning August 20th, 1852, and continuing, with little intervals, till the end of May, 1853.

On this occasion the first scoriæ were thrown up from the highest crater, after a violent shaking of the central nucleus of the volcano, and then for sixteen days continuous eruptions took place from two new cones, accompanied by two floods of lava, one of which ran two and a half English miles in the first eight hours. The whole descent of the flow was about 3500 feet. After a short lull further lava flows took place in various directions, precipitating in some places in a cascade of 400 feet, and not losing much of its fluidity after this fall. By this time the united breadth of the lava streams was not less than two English miles, and the distance or length of flow six miles. The piling up of the lava, poured forth perhaps in a less fluid state, continued for several months. The depth of the stream varied from eight to sixteen feet where it could spread out, but in some places reached 150 feet. Much of this lava was poured over the slopes of the cone, and remained standing at a high angle, as will presently be explained.

The circumstances connected with this lava flow were particularly favourable for determining the angle at which lava recently poured over a vertical face of rock, and also over rock inclined at various angles would present itself. Sir C. Lyell, on his visit in 1858, found that the surface of the continued sheets of hard, solid lava was inclined at angles of 35°, 40°, 45°, and in one spot 50°.

So far, also, was this from being an exceptional case, that he also found slopes of compact lava of older date dipping 35°, and others varying from 26° to 40°, partially covered by newer lava, also stony and compact, dipping 26°. Many other parallel cases are described in the Memoir by Sir C. Lyell already referred to, and it is evident that in Mount Etna such phenomena are by no means rare. It is also stated by Mr. Scrope that he himself saw lava streams hardening on Vesuvius at an angle of 33° in the years 1819-22, while Mr. Darwin, in describing the volcanic rocks of the Galapagos, and Mr. Dana, in the "Geology of the United States Expedition," both describe lavas almost vertical, evidently in the position in which they were thrust out of the earth. The latter traveller also mentions in the Sandwich Islands, "one overflowing layer of lava covering another, and forming a cone composed almost wholly of beds inclined at an outer slope of from 20° to 40°."

Without multiplying examples of this kind, which abound in all volcanic districts, we may take it for granted that the assumption of an elevation from below to produce the usual phenomena of a volcanic mountain is altogether unnecessary, nor will it explain at all the actual conditions of many known lava currents. It may be concluded that the high angle at which lava beds are often seen, assumed at one time as sufficient cause for the rejection of any theory but that involving upheaval, may be explained on other hypotheses, and thus the way is paved for more rational views of the subject; and we may proceed to discuss the second objection to the simpler hypothesis—namely, the nature and structure of the larger craters.

All those depressions that occur on the top or sides of a volcanic cone, whether deep, caldronlike cavities, cuplike depressions, or open, saucerlike pans, are known as craters, all having certainly been produced by the same agency—the outbursting of a quantity of volcanic ash, pumice, and scoria, with many large blocks of stone mingled generally with large volumes of water or aqueous vapour. The two theories of volcanos differ essentially in their explanation of the origin of craters, that which assumes paroxysmal elevation ascribing the depression, walled round on all sides with harder rock, as the result only of an elevation from beneath breaking the surface up, and leaving a rim around, just as if a thick bubble were formed and broken in some tenacious mass, and had burst on the convex surface. The explosion of a mine and its results are believed to afford the best parallel to the ordinary phenomena.

In some great eruptions there seems no doubt that the whole top of a volcanic mountain has been in a wonderfully short time destroyed and converted into a vast chasm by the sinking in of part of the mass. It may well be the case that a cavity is formed by the great and rapid movement which results from the first relief obtained by the forces acting below when the crust of the earth is broken, even without the cavity being the result of elevation. There are, indeed, too many and too distinct proofs of the formation of craters by the exhaustive process of a continued eruption to doubt that this latter is the real and forming cause. As Mr. Scrope observes, a single violent eruption, a mass of shattered rocks, and the sudden production of a crater, unfollowed by any other phenomena, is an event totally unrecorded, and is contrary to all experience. A volcanic eruption may, and often does, commence with noisy and repeated explosions, and much mechanical disturbance; but these are only the beginnings of the event. They are followed up quickly, or are accompanied by jets of vapour driven with immense force, rising many thousand feet. Mr. Scrope, describing what he saw of the commencement of the great eruption of Vesuvius in 1822, illustrates very clearly the subsequent phenomena, and points plainly to the probable cause of the peculiarities of form of the crater, and the grand result of the eruption. Without altogether quoting his account,

we may give the following abstract nearly in his
own words :—

The eructations of steam were accompanied by
scoriæ and fragmentary matter, several thousand
feet high, during a period of twenty days, and at
the end of that time had drilled through the
mountain an abrupt circular chasm three miles in
circumference and more than one thousand feet in
depth. The mass of matter thus removed, and a
large portion of the external summit of the cone
(which lost six hundred feet of height during the
eruption) had been blown into the air with the
steam, which rose to a still greater height than the
fountain of solid matter (at least ten thousand feet).
Each puff of vapour evidently consisted of the
contents of a great bubble, which had risen up
through the molten lava in the chimney of the
volcano, and burst on reaching its surface. To the
equal pressure in all directions of the enormous
expansive force of these flashes of steam may be
attributed the circular section of the crater or canal
of discharge, gradually bored through the heart of
the cone—continuous discharges taking place from
greater and greater depths, as the surface of the
boiling lava fell within the vent. By degrees the
explosions diminished in force and frequency, and
at length the tension of the vapour-bubbles bursting
at the bottom of the crater seemed to have no
longer power to throw off beyond its rim the
fragments which fell into it. The accumulation of

these at length choked the explosions, and the eruption terminated. In this eruption the fine ashes were not carried so as to form a deposit more than fifteen or twenty miles, even in the direction of the wind, but vast quantities were washed down the sides of the cone in streams of mud, produced by the condensation into rain of the vapour erupted. The coarser scoriæ and fragments, some of which weighed several tons, fell in abundance on the flanks of the mountain ; the average depth within a radius of five miles being not more than a foot or two. What must have been the force of some of the great eruptions on record compared with this, when we find ashes thickly scattered to a distance of seven hundred miles instead of fifteen, and an area of twenty-five miles radius covered ten feet after the eruption of Coseguina, in Central America, in 1835, or Sangay, also in Central America, in 1842-3, whose black ejected ashes covered the surrounding country to a distance of twelve miles, in beds from three to four hundred feet thick. But even these are not the most striking examples to be met with. There is the eruption of one of the Quito volcanos in 1797, when the mud, composed of ashes mixed with snow, filled valleys many miles in length, a thousand feet wide, to a depth of six hundred feet. An eruption took place in the Island of Sumbawa, in 1815, which continued for four months without interruption, throwing scoriæ and ashes in such abundance that they broke down the roofs of houses forty miles distant, and were carried more than three hundred miles in sufficient quantity to completely darken the air at that distance, while the floating cinders in the ocean formed a mass two feet thick, through which ships could hardly force their way. Doubtless the great paroxysmal eruption of Vesuvius, in the year 79 of the Christian era, was of similar character, but the accounts of it handed down prove clearly that it was a violent eruption of the ordinary kind. A vast portion of the old cone was then blown away— the deposit of ashes at Naples was then probably from six to ten feet, and three populous cities were buried under from fifteen to one hundred and fifty feet of erupted matter. Mr. Scrope well remarks that a due consideration of what must have been the size of the hollows (*i.e.*, the craters) left by the forcible explosion of these startling quantities of matter, from the centre of a volcanic mountain, will account for the largest known craters, without any theory of the subsidence of mountain-tops, or the circular upheaval of previously horizontal beds.

The whole principle is involved in the explanation of these phenomena. That vast eruptions of the kind alluded to do take place, with comparative frequency, in various parts of the earth, no one has attempted to deny; and it is certainly more reasonable to assume as the real cause of the hollow, the known fact of the removal of the whole interior mass of matter which had been gradually, and for a long time, accumulating there in a melted state, than to suppose that the summit of some vast empty gulf, assumed to exist below, has fallen in, swallowing up a large part of the mountain.

There is, indeed, another kind of crater, of which a grand example has been described in the Sandwich Islands. In this case a large quantity of liquid lava boils continually from a vent gradually raised by the cooling of the lava on its edges, occasionally overflowed by the lava rising above the edges, and from time to time topped, as it were, by an earthquake movement, splitting open the brittle sides or walls that contain the lava, and allowing it to escape to a much lower level. A circular or oval cavity is thus formed by subsidence, and the result is a crater surrounded by perpendicular cliffs forming a series of steps, each marking a stage of the progress.

That many parts of volcanic districts are worn away by the action of water, and that sometimes deposits containing fossil shells are found at a high elevation on their cones, has been regarded as an argument in favour of the upheaval or paroxysmal theory.

The removal of enormous quantities of volcanic matter from where it was originally heaped, and the action of water in this movement, have been recognised, especially in the Val del Bove, immediately adjacent to the great cone of Etna, and this has been attributed to a great and sudden catastrophe connected with movements to which the steep outward dip of the lava beds on the flanks are supposed to have been due.

In a Memoir recently published,* Sir C. Lyell has very carefully considered the structure of Etna, and has succeeded in showing the high probability of a double axis existing, the older of which passed through the Val del Bove, where the more ancient eruptions occurred, while the later eruptions have taken place through the newer axis under the existing cone and crater.

The idea of such double axis is not new, nor has it been invented to support any hypothesis. From the observations of certain dikes of hard greenstone, which consist of the lava that has been last poured out, and has remained in a crevice after a large quantity has flowed through it, the Baron S. Von Waltershausen (the author of a most elaborate map of Etna, and an accomplished observer of volcanic phenomena) had before inferred its existence, and had pointed out its position, and other geologists have since confirmed the probability of it. It is also an inevitable result of all modern investigations into the history of Etna, that the Val del Bove was not formed till long after the whole mountain had been formed with its lava and tuffs, and had been pierced with a succession of dikes marking the direction of old lava currents.

* Philosophical Transactions of the Royal Society (for 1858), vol. cxlviii. p. 703.

Still the existence of the grand valley, of which one of the bounding cliffs is nearly four thousand feet high, whose width is from two to three English miles, and length more than ten miles between the foot of the principal cone and the sea, is a phenomenon of too large a magnitude to be very readily explained. Sir C. Lyell believes that a large part of it may be the result of the floods that pour down the flanks of the cone; but that a direction had been given to these torrents of water, perhaps by cracks produced during some of the paroxysmal eruptions without much lava. Of such crevices the eruption of the year 79, already alluded to, affords a striking example.

However this may be, there seems nothing in the Val del Bove to justify the assumption that there has been in any part of it more than a small elevation corresponding to that noticed on the adjacent shore; and it is well known that a slow process of upheaval has been going on for a long period over all this part of Europe. There is, however, another argument of some ingenuity which completely sets at rest the question as to the sudden formation of Etna at any date by sudden and paroxysmal elevation. A multitude of dikes traverse the material of which all the cones are made up, and these abound at the two principal cones near the assumed axes. These dikes, from the nature of the case, are more or less highly inclined, and are often vertical, but their position in regard to each other, and to the beds they traverse, renders it impossible that they can be contemporaneous.

It may be concluded, then, that although a very considerable though partial elevation has taken place in parts of Etna on the eastern and southern base, these commenced at a comparatively recent geological period, and no connexion can be traced between this gradual movement and the conical or domelike form of the mountain; and even where the strata containing shells and animal or vegetable remains have been burst through by local eruptions, they have not been lifted up in such a manner as to favour the hypothesis of craters of elevation.

It yet remains, indeed, to be seen how far the form, whether of conical or dome-shaped volcanos, may be owing to a gradual distension of the mass brought about by the injection of dikes of melted lava, thus giving to the outer beds a steeper inclination than properly belonged to them; but these causes are not sufficient to account for more than a slight increase in dips already far more considerable than has sometimes been assumed as possible, in the case of semi-fluid molten lava poured out on a hill-side.

We may conclude with one more argument, not without its value, tending to show how little in accordance with probability is the idea of a paroxysmal elevation of volcanos. It is a view of Mr. Darwin's that "there is local antagonism rather than coincidence between direct elevation and volcanic action, and that in fact dislocations on a large scale are rare in volcanic districts." Nor is this unreasonable, for if the earthquake is the shake and tremor produced when an impulsive force endeavours to find vent and reach the surface; it is reasonable to suppose that if the force does find vent, there should be no more disturbance, and that such force should not be chiefly exerted when the vicinity of a former vent renders the production of a new one comparatively easy. In either way it may be assumed that the volcanic eruption has first taken place when the earthquake has rent the earth, and that the facility for eruption afterwards prevents that considerable and permanent upheaval which would be required for the production of a new volcano on the "elevation-crater" principle.

Should it appear that all the principal volcanic cones have been produced by the comparatively slow process of eruption repeated from time to time at intervals of many years, we shall come to the conclusion that the period required for the production of those volcanos that are best known is very considerable, and must admit of great changes in the level of land, great climatal changes, and even modifications of the animal and vegetable tribes inhabiting land and water. Most of these volcanos, if not all, are, however, of comparatively modern origin, and they rarely seem to carry us back even to what has generally been regarded as

the Last Tertiary as distinguished from the Recent
Period. Many, again, of those which now seem
permanently quiet, such as those of Central France
and the Rhine, were certainly active during the
Tertiary Period, whilst the great lava currents of
the north-east of Ireland and its neighbourhood,
seen in the Giant's Causeway and the Isle of Staffa,
are newer than the chalk. It would seem, therefore, that the life of a volcano or volcanic region is
not of long duration in comparison with many
geological events.

IX.

THE BATTLE OF LIFE.

A struggle for existence the great law of nature—Illustration of the different kinds of struggles in animal and vegetable life—Why a species abounds in any locality—How it is affected by change—How change is produced — Example shown in the breeding of sheep—Effect of crossing breeds—Effect in pigeons —Exceptional case of mules—Views of the introduction of species — Mr. Darwin's idea of nature's method that of Selection—Meaning in which this method is understood— Selection a principle long known—Limit of human observation and power in the application of this method—Gradual modification or extinction of species inevitable—History of a supposed change from one species to another—Limits supposed to mark natural species not real—Slowness of nature's process of change—Further illustration of "the Great Battle."

THAT existence is for all plants and animals an individual struggle—that all are exposed to the attacks of enemies from earliest youth to extreme age—and that in a state of nature none but the strong, healthy, and energetic individuals come to maturity under ordinary circumstances—these are facts too well known to need illustration. That whole species or natural groups of animals and plants are subject to the action of laws closely resembling those which govern individuals is a fact which most naturalists are prepared to admit, although general readers in natural history rarely pay attention to questions apparently so technical. But that the constant struggle for existence and mastery thus affecting individuals and species should result in a series of modifications producing what naturalists call " permanent varieties," equivalent, in fact, to distinct species, is, if not a new idea, at least an idea so pregnant with important consequences, if fairly worked out, as to justify the closest attention both to the assumption itself and to the legitimate deductions from it.

The Battle of Life—that struggle for existence which limits the numbers and distribution of all living things upon the earth—is undoubtedly a proper study for every one; and at the present time, when geologists are rapidly extending the basis of their science, and assuming that they have already obtained sufficient facts for broad generalizations in natural history, it is a study of especial interest and importance.

There are many aspects in which the battle of life may be viewed—many kinds of struggles in which all living things are engaged. A dog and a wolf, both inhabitants of a country of limited extent, may in time of dearth have to struggle with each other which shall live. The dog may be the more swift, and in a general way more intelligent; the wolf stronger, and possessed of more low cunning; it will depend on the nature of the prey obtainable which shall live and which starve. It may be that the dearth continues, and the whole question of the existence of the race may be connected with the struggle. If the last remaining prey is that which will be gained by the dog, the whole breed of wolves must perish from the locality, as has happened in our own island, although with us the intervention of man and the domestication of the dog have doubtless assisted. In the case of the rat and the beaver, however, this is hardly so, and as the beaver has long ago been driven out and the rat still remains, the struggle has here terminated in the extinction of the more intelligent and larger animal, and the increase of the bolder, fiercer, and smaller kind. This is one struggle in which, although the progress of a dominant race may have indirectly aided, a great battle has been fought between allied races, and a complete victory gained by one over another.

There is another struggle less obvious and literal, as when a plant on the edge of a desert, dependent on moisture, battles not so much with other plants as with the sands of the desert; as long as the plant grows the advance of the sands is checked; as soon as drought destroys the plant, the sands occupy the ground, and the possibility of existence of the plant is at an end.

Again, of the hundreds of seeds which a plant produces, but very few come to maturity, and of these not many germinate under ordinary circumstances, there being a constant and incessant struggle for the ground, the nutriment derived from the soil, and the moisture, with hundreds of other plants of the vicinity. For a long time when the balance is once struck each plant remains and occupies its place, until another plant, somewhat 'more fitted to take advantage of the soil and the climate, is introduced from a neighbouring land, or from the antipodes, and this new plant may then entirely take the place of the old one, and drive it out. The old inhabitant is exterminated as the Red Indian and the Australian are giving way before the advance of the Saxon races of men. In a similar way the recently introduced horse and pig have in many parts of the world exterminated native animals.

In the case of every animal and plant there are far more individuals produced than can be reared; and if any one of them were to continue its increase according to its own natural rate without check, it would soon stock the world. Linnæus calculated that if an annual plant produced only two seeds, and their seedlings next year two, and so on, in twenty years there would be a million plants. If the elephant—the slowest breeder of all known animals—be assumed to breed when thirty years old, and to go on breeding till ninety years, bringing forth three pair of young in this interval, then at the end of the fifth century there would be fifteen millions of elephants alive all descended from the first pair.

So abundant and vast is the increase, even where the smallest number of seeds ripen, of eggs are hatched, or of young are born, that in any calculation of the number of individuals that may be expected after a certain interval under given circumstances, the rate of increase, the number of seeds, or eggs, or the frequency of pairing is altogether a secondary consideration and comparatively unimportant.

"In looking at nature it is most necessary to keep the foregoing considerations always in mind, never to forget that every single organic being around us may be said to be striving to the utmost to increase in numbers, that each lives by a struggle at some period of its life, that heavy destruction inevitably falls either on the young or old during each generation or at recurring intervals. Lighten any check, mitigate the destruction ever so little, and the nnmber of the species will almost instantaneously increase to any amount. The face of Nature may be compared to a yielding surface with ten thousand sharp wedges packed close together, and driven inwards by incessant blows, sometimes one wedge being struck, and then another with greater force."*

Such are the words of Mr. Darwin in reference to the struggle for existence quoted from his work recently published "On the Origin of Species," in which this cautious, learned, and accurate naturalist

* Darwin's "Origin of Species," p. 88.

has given the result of nearly a quarter of a century's inquiry, thought, and observation. Without at present alluding to the great object of his investigation, let us follow out some of the results of the great struggle in illustration of the title of this article.

A species of animal or plant abundant in any particular locality is so because circumstances are favourable for its development, and are not so favourable for others at hand that would replace it if they could. There is, however, no permanence in existing things — for no two seasons are exactly alike—cold and heat, wet and dry, shelter and exposure all vary from year to year, while the slightest change in almost any of these may in a thousand indirect ways affect any species. Every species, therefore, without exception, must be subject to occasional crowding out, producing starvation, if it be not to some extent capable of adapting itself to changing circumstances. If it should not be thus far capable of change, either in itself or in some of its offspring, it can only be abundant for a short time, and will then be lost altogether. If it be changeable in any important respect, or if of the rising generation of plants or animals of any species, some individuals are more readily altered, or are naturally more modified in a favourable direction than the rest, then there will be the commencement of a variety, formed by the accidental peculiarity of some one member of a group. Owing to the well known law of resemblance of the offspring to the parent, there will probably be some of the next succession who possess this peculiarity of the parents whatever it may be. Out of the whole number, then, those which are strongest and best able to fight their way must succeed, and the rest fail and die, being beaten in the battle. If therefore the peculiarity is advantageous, it will be perpetuated, if unfavourable, it will be lost.

An excellent illustration of this method is seen in the history of a race of sheep, valuable for their wool, whose first origin traces back to a male lamb of the merino breed, belonging to a farmer at Mauchamp, in France. As it grew up, this animal, though small and of inferior general conformation, was found to possess a peculiar silky character of wool, which rendered it specially valuable for the manufacture of Cashmere shawls. By breeding from this ram the farmer obtained at first two lambs (a ram and ewe) possessing the peculiarity. From these he got others, and in the course of time a distinct breed, which may be regarded as a permanent variety. By degrees and by careful management, the defects of the original parent were got rid of, and the race now has the silky wool without the least deterioration, combined with symmetrical form and the ordinary size of sheep.

It becomes then a most important inquiry how far this production of a variety can go, and what it leads to. As far as man is concerned, he can only

take advantage of what he sees, and his selection
of peculiarities from which a permanent variety
can be secured is confined to a few external characters.   No doubt these are generally more or less
directly connected with structure, but the extent
of this relation is rarely made out, and they are,
also, almost always confined to domesticated animals
and cultivated plants ; which form of necessity but
a small number of all animals and plants.   The
clue to nature's method, however, must be sought
for in the results of experience in gardening and
breeding.

There are some curious rules or laws of nature
with regard to this subject; one of the most
curious and best known being, that crosses or
mixed breeds, when obtained from animals or
plants of altogether different kinds, are almost
uniformly sterile; but those obtained between varieties of what are regarded as the same kind, are
as uniformly not only fertile, but more fertile than
the breeds obtained from individuals more nearly
related.   Thus, from the horse and ass we obtain
hybrids (mules) which are always sterile, but from
the mixture of different breeds of horses, cattle,
dogs, and sheep, we get mongrels, which are generally perfectly fertile, and often more so than individuals of the same variety are when bred together.
At first this appears to prove that to obtain a
permanent fertile variety, altogether different from
the original parent, is a natural impossibility, and

such has been the general opinion of naturalists. There is reason to believe that this conclusion, however apparent, is not altogether correct, and that nature really does of herself produce animals extremely different from those existing in a given place at a given time, provided only sufficient time be allowed her to produce the required change by the slow steps she has arranged. Not only are such modified forms to be found in nature, but man is enabled to imitate nature's method to some extent, and procure them almost at will by proper management. It will be evident that, whether species are rigidly fixed and permanent, or admit of wide modifications in the great struggle for existence, is a question whose determination is of the very first importance.

In the work already mentioned, Mr. Darwin illustrates the variability of natural forms of animals and plants in many ways, and by reference to various so-called species. He set himself to work upon pigeons, these birds being convenient for many reasons, and derived as distinctly as any group of varieties can be proved to be from one common parent—the rock pigeon of Europe. The differences now observable amongst these birds are so very great and so important, that the example is one well adapted to illustrate the position he was anxious to prove — namely, that the varieties obtainable from a known species involve diffe-rences positively *greater* than those universally agreed on as distinguishing species amongst allied animals.

Thus, in the skeletons of the various breeds, the relative length and breadth of the different bones of the face differ enormously; the number of vertebral bones of the tail and back vary; the number, relative breadth, and attachment of the ribs is different; the shape, length, and breadth of the branches of the lower jaw vary extremely. These and other differences in the skeleton are also much more considerable than are found in birds regarded as of altogether distinct species. Externally, we find the proportionate dimensions of the gape of the eyelids, of the orifice of the nostrils, and of the tongue, the number of the primary wing and tail feathers, the proportions of the wing and tail to each other, and to the body; the proportion of the legs and feet, and the form of the little scales on the toes, all points of structure which are variable. The time when perfect plumage is acquired, the shape and size of the eggs, the habits of nest-making, the manner of flight, the voice, and the disposition also vary. As a general result, it may be said that at least a score of well-defined species might be made and grouped into several genera from what almost all naturalists are prepared to admit to be varieties of one species (the rock pigeon), which is the only wild species that has the same habits as the domestic breeds, and to which all the domesticated kinds show their relation by their peculiar tendency to return to certain points of colour and marking which are characteristic of their wild ancestor. It is worth mentioning, as some explanation of the extraordinary amount of variety shown, that these birds have been domesticated at least five thousand years, as is proved by historic records, and that in the time of Pliny, the luxurious Romans gave immense prices for certain varieties, and reckoned their pedigree and race.

What has been said for pigeons might be repeated, with some variation, for dogs and horses, for sheep, and for horned cattle, the breeds of each of which were probably obtained from some one, or very few, parent species. Certainly the differences between the greyhound and the turnspit, the Blenheim spaniel and the Newfoundland dog, the dray-horse and the race-horse, the Shetland pony and the Flemish coach-horse, the Scotch and Durham breeds of oxen, and the various breeds of sheep, are as marked as those between the different kinds of pigeons; but in all these, there is every reason to suppose that the change has been wrought during domestication, by the agency of man. How is it, then, that man can act so as to produce modifications of form, structure, and habit so complete. He cannot, certainly, fight against nature, for he cannot obtain a fertile cross-breed between any two species that are extremely different. The sterile mule is almost the only really important hybrid that has ever been obtained by the in-

genuity of the human race, and there has certainly
been no failure for want of endeavour. The fact seems
to be, that as in nature there are no sudden tran-
sitions, man must be content to study the course of
nature herself, and imitate it, or, failing to do this,
he must work out the problem on his own account.
He will generally find, when he has done so, that he
is really only availing himself of a law he might have
discovered by a simpler process. But this law, till
lately, has not been properly set forth, though very
frequently acted on, and some even philosophical
naturalists, as well as some popular writers, have
imagined that an animal or plant, out of the mere
want of some organ, and, as it were, from a longing
desire to possess it, has at length, during a succession of generations, attained to it. Others, offended
at a theory so opposed to reason and probability,
have assumed new species to be suddenly and from
time to time born from some one previously existing,
by a miraculous intervention involving an immediate and extreme variation at one step. Others again
have assumed occasional independent creations of
new species in large groups. Every person who has
studied nature, and still more every one who has pursued even very superficially the arguments of geology, must have felt the introduction of species to be
an enormous difficulty, rarely approached and never
fairly met.

Without pretending to dogmatize in this obscure
department of science, we may perhaps assert that

the clue recently given by Mr. Darwin, in his " Origin of Species," if not actually to be depended on as affording a complete elucidation of the mystery, is yet extremely valuable, and leads to great and important truths in that direction. The principle is stated in one word — SELECTION. Studied first in its action on domesticated animals in producing the great amount of variety barely outlined in what has been said above, this great principle may be shown to be generally acted on by nature. It also expresses the method by which, in the great battle of life, the balance is everywhere produced, and that never-ceasing natural variety preserved which is one of the most striking illustrations of the infinite power and wisdom of the Original Designer. It ought not for a moment to be supposed that man was the first to take advantage of the wonderful adaptability of all created things. It is so manifestly *the* law of nature in the development of life, that, once enunciated, it seems like the case of Columbus and the egg, and every one exclaims that the idea is too simple and manifest to deserve the name of discovery. Discovery, in one sense, it may not—perhaps cannot—be; but in its application to solve the great mystery of the gradual modification of old and the production of new species, it seems to us that Mr. Darwin deserves all the credit that belongs to one who has thoughtfully, and with great labour, investigated a large group of facts, and indicated their meaning. In connecting these facts, he has not only been the first to see the inevitable result, but has been honest enough and bold enough to state his conclusions in distinct language, not concealing the difficulties or objections, and not pretending to explain away all the first or satisfactorily reply to all the latter.

The "selection" alluded to by Mr. Darwin is described by Mr. Youatt (well-known for his acquaintance with the peculiarities of breed in animals) as that principle "which enables the agriculturist not only to modify the character of a flock, but to change it altogether. It is the magician's wand, by means of which he may summon into life whatever form and mould he pleases."

Strong as this language is, it is hardly exaggerated, and is, indeed, fully proved by the most recent results of breeding in all departments.

Nor is this done by crossing different breeds, and taking a variety accidentally thus obtained; for, if selection consisted merely in this, "the principle would be so obvious as hardly to be worth notice; but its importance consists in the great effect produced by the accumulation in one direction, during successive generations, of differences absolutely inappreciable by an uneducated eye. Not one man in a thousand has accuracy of eye and judgment sufficient to become an eminent breeder. If gifted with these qualities, and he studies his subject for years, and devotes his lifetime to it with indomitable perseverance, he will succeed, and may make great improvements: if he wants any of these qualities, he will assuredly fail."*

This principle of selection is by no means a modern invention as applied to plants and domestic animals. Certainly Jacob took advantage of it in providing for himself out of the flocks of his father-in-law, as described in Genesis xxx.; and the method is quoted in an ancient Chinese cyclopædia, while explicit rules based upon it are laid down by some of the Roman classical writers. After all, however, man can only perceive outward peculiarities, which may, or may not, have an essential bearing on the structure of the animal; and thus all his efforts are subject to disappointment. He endeavours also to obtain some peculiarity which very often is not for the real advantage of the animal, except in its reference to some special use. Nature is at no such disadvantage; and if, as is abundantly shown by the efforts of breeders, animals and plants generally are capable of enormous variation, if the change be effected by selecting for breeding or growing those individuals who possess an indication of the required peculiarity, carefully watching the process from generation to generation, we may understand that when a favourable variety is produced naturally, it may also be continued and improved into a permanent variety in a state of nature.

---

* Origin of Species, p. 32.

What is here meant by a favourable variety is one
that, under the existing circumstances, whatever they
may be, is more able to oppose its enemies, and
obtain abundant proper food, than other varieties, or
than the original stock. " Owing to the struggle for
life," as Mr. Darwin says, " any variation, however
slight, and from whatever cause proceeding, if it
be in any degree profitable to an individual of any
species, in its infinitely complex relations to other
organic beings, and to external nature, will tend to
the preservation of that individual, and will generally be inherited by its offspring. The offspring
also will thus have a better chance of surviving;
for, of the many individuals of any species which
are periodically born, but a small number can
survive."* This is the principle of *natural
selection*—a power incessantly ready for action, and
as immeasurably superior to man's feeble efforts as
the works of nature are to those of art.

The struggle for existence—the battle of life—
is the great guiding cause of natural selection, and
is going on around us, and before us, in every
department of nature. Every plant and every
animal tends to increase with enormous rapidity.
In an old-established country each one is kept
within certain bounds by a system of natural
checks. These, however, are not permanent, for
although we think there is no change, there cannot

* Origin of Species, p. 61.

be a doubt that even periodical modifications of weather involve alterations in the distribution of all living races, that may have permanent influence on their organization. Where, however, by any accident, or by the hand of man, a new plant or animal adapted to the climate has been introduced into an island or newly-discovered continent, without its natural checks, it generally increases with enormous rapidity, destroying the indigenous tribes. Thus, several of the plants now clothing square leagues of surface, almost to the exclusion of all other plants throughout the wide plains of La Plata, were introduced from Europe; and there are plants in India, ranging from Cape Comorin to the Himalaya, which have been imported from America since its discovery. So, also, the plains of South America and Australia are covered with cattle and horses recently introduced. The only limit in cases of this kind, where the natural checks are absent, is the supply of food; but in time no doubt other races will arise and reduce these newly-introduced ones within narrower limits. The number and variety of the mutual checks to indefinite increase, both of plants and animals, is altogether unlimited; and these checks act at different periods of life—some chiefly at one season, and some at another—some are potent for a period, and then powerless; while others are ever at hand to replace any that are lost.

If, now, we take the case of a country under-

going some physical change, as of climate, we
shall see that, following the inevitable laws of
nature, some great modifications *must* take place in
the living tribes, to provide against their extinction. If, for example, in such a case, the deer
increased in numbers, and other prey of the carnivora decreased, the deer being the fleetest, the
carnivorous animals whose proportions and habits
are best fitted to pursue and take such prey would
have the best chance of surviving, and would
therefore inevitably survive the rest. Thus, the
swiftest and slimmest wolves would have an advantage over the clumsier individuals; and a breed of
swift and slim wolves would therefore be in time
obtained by nature, just in the same way that man
would succeed in improving and rendering more
fleet a race of greyhounds by careful and metho-
dical selection, if he desired to do so, and chose to
take the proper means. This, however, is only one
mode of natural selection; and it is quite possible,
and highly probable, that, as the cubs of beasts of
prey are often born with an instinct to pursue
certain kinds of prey rather than others, some individual might appear having the favourable peculiarity. Such cases are exceedingly common; for
it is well known that one cat will catch rats, and
another mice, while a third will bring home game,
and this in one case winged, in another hares or
rabbits—each according to its instinct and powers.
It is quite clear that, as instincts are hereditary as

well as points of structure, there might soon be a
variety of any species secured whose habits and
powers led it to pursue one kind of game rather
than another, provided any one kind were to be obtained more abundantly than others. On the other
hand, if one kind should become very rare or
extinct, then the wild species who preferred, or
could best obtain, that kind would either change
their food or themselves become extinct also. As
an example of this, it may be mentioned that two
varieties of wolf exist in the Catskill mountains,
in the United States—one kind shaped like a greyhound, and pursuing deer; while another kind is
more bulky, with shorter legs, and attacks sheep.
These certainly must be natural varieties produced
by natural selection.

More complicated cases of struggle and accompanying selection may easily be found. Mr.
Darwin mentions an instance of some plants which
occasionally excrete a sweet juice sought by insects.
In the commencement, when this juice is only
accidentally or occasionally excreted in the flower,
insects seeking the juice would inevitably take the
pollen from one flower and transport it to the
stigma of another belonging to another plant of
the same species—thus crossing the breed, and no
doubt in this way increasing the fertility, and
tending to produce vigorous seedlings, which,
simply because they are vigorous, would probably
excrete more juice. In the long run, those flowers

which excrete most freely, and which, by any peculiarity of position of the pollen, stigma and excreting ducts, are most favourable to crossing by the aid of insects, must prevail. Our imaginary plant being thus far advanced, any little tendency to the separation of the sexes, such as is by no means uncommon, would certainly be taken advantage of in consequence of the frequent visits of insects helping to carry pollen, and thus increasing the number of cases in which the separate sexes would be fertilized. Thus, those plants having only stamens, and those having only pistils, which would, according to a known law of nature, be the most vigorous, will very soon entirely drive out the hermaphrodites.

We have thus obtained from a plant containing both pistils and stamens, and only occasionally exuding nectar, another plant, with sexes distinct, always exuding nectar and necessary for the insects, which, while they fed on the nectar, assist in the propagation of the plant. By one further step— the form of the flower admitting of great modification, and certain shapes being most easily reached by the nectar-feeding insects—the most favourable shapes for these purposes would replace others and become most frequently propagated.

During all this time, the insect, perhaps a bee, at first accidentally feeding on the accidentally-produced nectar, must be undergoing a corresponding modification under a precisely similar law;

those individuals who possess any special aptitude, by instinct or structure, to feed on the nectar, being multiplied; and those which are otherwise developed either dying or turning to other habits, thus ending in the production of a race of insects feeding on honey, and of plants yielding honey, adapted perfectly to each other, and mutually dependent. If, after this, either should be cut off, the other must also suffer; and this also would clearly be a consequence of the natural working out of the same law which, if it exists, must act wherever circumstances permit.

The intercrossing first of individuals, then of families, and still further of varieties within certain unknown limits, is one of the laws by which natural selection is assisted, and one of the means of greatly increasing fertility. Beyond these limits, intercrossing is unnatural and sterile. It has always been a favourite speculation of naturalists that these limits, whatever they may be, mark natural species —that all fertile intercrosses can only take place between varieties, and that when sterility is the result, the species must be distinct.

Such, however, is by no means the only explanation of an undenied fact; for whenever varieties have separated so far from the original stock as to have produced races which, in proportions, habits, or any of those peculiarities which influence breeding, tend to interfere with due development of the young, sterility is very likely to be caused by

.the early (often premature) destruction of the offspring, instead of the absolute infertility of the
parent. Thus, at length, a time comes when the
descendants are so far removed from resemblance
to each other that they cease to propagate.

This removal would seem to require the lapse of
a long period of time—longer, perhaps, than is
accounted for by any continuous history recorded by
man, though probably by no means so long as is involved in what is called the historic period. Still
the process of selection by nature, though ultimately
extending much more widely and producing far
greater divergencies, will probably be slower than
that by human agency. Nothing will be done
unless favourable variations occur, and these generally must have reference to physical changes, which
are in their nature extremely slow. To allow of an
important change in organization, there must be not
only such physical changes, but the immigration
of other better adapted forms must be checked in
order to give the requisite time for the modification
of the existing races. There will also be long
intervals during which nothing occurs to render a
variation advantageous, and generally only a
few of the inhabitants of the same region will be
affected at the same time. No argument of this
kind, it may be remarked, can take away from
the value of natural selection, as a *vera causa*
accounting for varieties of specific form, and for the
introduction of those permanent varieties, which are

in reality as much species as any others. This is the case, because indefinite time (not in any way, however, approaching infinite time) must be an element granted in all speculations concerning the origin and succession of created things.

Once more, then, we recur to the title of this essay, "The Battle of Life," and venture to assert, as beyond contradiction, that it not only exists, and always has existed, in the strictest sense of the word, but that it is a very essential part of the great plan of creation, and leads to the recognition of one of the great laws of nature. The struggle commenced from the very beginning, and is a necessary consequence of the law of increase by geometrical ratio common to all organic beings. There is no known exception to this law, as every species tends to increase rapidly or it could not exist at all.

It is certain that of all natural families of animals and plants many " more individuals are born than can possibly survive. A grain in the balance will determine which individual shall live and which shall die—which variety or species shall increase in number and which shall decrease or finally become extinct. As the individuals of the same species come in all respects into the closest competition with each other, the struggle will generally be most severe between them ; it will be almost equally severe between the varieties of the same species, and next in severity between species

of the same genus.   But the struggle will often be
very severe between beings most remote in the
scale of nature.   The slightest advantage in one
being at any age or during any season over those
with which it comes into competition, or better
adaptation, in however slight a degree, to the
surrounding physical conditions, will turn the
balance."*

The struggle will often be that of males for the
possession of the females.   The most vigorous
individuals will generally leave most progeny, but
success will often depend on the accidental possession of special weapons or of special charms, which
may also be in one sense accidental.   These also,
however, will be transmitted and become the
foundation of varieties—often gradually introducing
wider divergences in the same direction.   The spur
of the game cock and the tail of the peacock must
be regarded as examples of these two causes of
victory.

Generally, however, new and improved varieties
will supplant and exterminate those that are older
and less improved, and so ultimately species are
formed dissociated from others, having different
objects.   Dominant species tend to produce new
and dominant forms, each large group becoming
larger and more divergent, and as all cannot
advance, the more dominant beat, and ultimately

* Origin of Species, p. 467.

destroy, the less dominant. In this way we can best understand the grouping of all organic beings within a few great classes, which have always been the same, and which have mutual and very definite relations.

Thus, then, the inevitable result of the incessant fight is to produce order and system in nature, to ensure the keeping up of every race by the individuals best adapted to the circumstances of the day, whatever those may be, and to preserve in perfect harmony and well-being the grand scheme of creation, according to which life is everywhere present and is always tending to higher forms of development and greater complexity of organization. Surely there is nothing in this view of the method of creation that can be regarded as derogatory to the power and dignity of the Great Creator. The gradual derivation of species from varieties, produced according to the action of a permanent and unbroken law imposed on organization, is as great an exhibition of power as the occasional infraction of a law of partial extent, and the constant recurrence of a special act of creation. It is not necessary, nor is it desirable, to follow out here the logical deductions of the great law enunciated, or to suggest the difficulties that surround this, as well as every other attempt that has been made to explain the constant introduction of new species throughout geological time. It is enough that the laws of variation of species and of the inheritance of variations, both of structure and instinct, are absolutely proved by the careful observation and invariable experience of civilized man from the earliest recorded time.   The existence of such laws being granted, it is utterly unreasonable to suppose that they apply only to the races of animals accidently domesticated, or that they should have lain dormant till brought into action by man.   If they are laws of nature, they cannot fail to have produced infinitely greater and more perfect results during the long periods of geological time, and in the grand physical changes that have taken place in that time, than can have been the case during the comparatively short duration and small extent of human change.   The principle of natural selection once admitted and fairly applied to the circumstances of the great struggle for existence, there is no need to answer in any way those who will endeavour to reply to observation and investigation by ridicule, for the principle and the argument will work their own way, and cannot fail to lead those willing to learn to discover ultimately the real method adopted by nature in the introduction of new forms of vegetable and animal life upon the globe.

X.

ANTIQUITY OF THE HUMAN RACE IN EGYPT.

*Geological errors—Hasty assumption of theory a fertile cause of
such errors—Progress of discovery in science—Human remains
with fossils no new discovery—Refusal to admit the evidence on
this subject—Inquiry suggested by Mr. Horner in Egypt—
Fitness of Egypt for the investigation—Delta of the Nile—
Selection of spots for boring into the mud of the Nile Delta—
Estimate of the rate of deposit of the Nile mud—Result of
sinking, and discovery of human remains—Consequent great
antiquity of the human race in Egypt—The Chronology of the
Bible an open question.*

AN account of the correction of mistakes in geology
might furnish matter for many very amusing and
instructive chapters in a work like the present. Few
of the younger geologists of the day, and fewer
still among general readers, have any idea of the
extent to which opinions have become imperceptibly
modified in many important departments of geological science within the last quarter of a century;
while there have not been wanting several absolute
and formal recantations enforced from time to time
by direct discovery. The great cause of this is to
be found in the inveterate habits that almost all of
us have of over-estimating the value of negative

evidence. Geologists examine, as they think carefully, a certain district or a certain collection of
objects, and remark the absence of some object or
group concerning which there seems no good
reason why it should not have been handed down
at least as perfectly as some others that have been
preserved. At once the theorist jumps to the conclusion that the tribe of animals not represented
had not yet been created. It is an easy and com-
fortable conclusion. A theory is soon built up on
the strength of it, for no one can oppose it without
having the *onus probandi* thrown upon him ; and
if it is difficult to prove a negative, to disprove
one without an actual positive fact at hand is still
more difficult. But some fine day the required
fact is discovered, often to the disgust of the
theorists—to the equal vexation of the student,
and it would almost seem to the annoyance of everybody. The first impulse of human nature is to
put the unlucky discovery on one side—say nothing
about it—most likely it will not bear investigation,
and therefore don't let us have the trouble of
investigating it. Such is the impulse of even the
honest, scientific, hard-working naturalist. It is
so painful to be stopped in a pleasant career of
progress, and to be obliged to examine carefully,
and weigh fairly, the evidence in regard to a
matter we thought settled when we began work
some twenty years ago.

But the progress of science is made up of steps
of this kind—first, many forward steps are secured ;

after a time, some of those thought secure are lost, and we must struggle again; but the result is, after all, real advance, and no one should be discouraged at so natural a law. Certainly geology is not the science least troubled with discoveries of this kind, for there are few in which, at the outset, the pursuers took longer steps, or seemed to rise more rapidly from elevation to elevation. There are, perhaps, not many in which so much has to be unlearnt.

A quarter of a century ago the minds of geologists were quite made up about the question of human fossils, or the existence of any remains indicating a longer duration of the human family than the few thousand years attributed to it by the prevailing chronologies. No fossil men were found— no human bones were mixed up with those of the remarkable animals that seemed to have lived on the earth but lately. Some apparent exceptions were easily explained away, and the general conclusion was considered as established. Man was the last animal introduced on the earth, and, compared with any geological event, he was a creature of yesterday. As this view happened to fall in with a recognised interpretation of the Sacred Record, it was the more willingly accepted, and was likely to be the more firmly retained. More theories and general views have been based on this supposed logical conclusion than we should care to enumerate, but—

" Mark how a plain tale shall set him down."

For some years after the publication of Buckland's "Reliquiæ Diluvianæ," all seemed smooth enough. But, even at that time, a troublesome Frenchman—a M. Boucher de Perthes, of whose work we shall speak more at length under another heading—took it into his head that some remains of men ought to be found in gravel. Now, true gravel was a domain into which man, according to the orthodox conclusions, never could have been admitted. Lapse of time, change of climate, destruction of important races—all rendered it impossible; so poor M. de Perthes, although he found plenty of specimens, figured some hundreds of them carefully, published an octavo volume about them, and even offered his specimens to the savans of Paris, could not obtain a hearing. Few readers, either in France or England, seem even to have been aware of his book, although one published a translation of part of it at Liverpool, which had the same fate as the original. The subject was tabooed, not with any real wish or idea of suppressing truth, but because people's minds were quite made up on the subject, confiding in the strength of the negative evidence, which really meant little more than a total absence of inquiry.

At length, however—for "*magna est veritas et prevalebit*"—something like what had really been found in France twenty-five years ago, was found in England. The result of this discovery, and the investigation that ensued, will be found recorded in another chapter; but meanwhile, another kind of evidence was introduced in consequence of an interest taken by Mr. Leonard Horner in boring into the alluvial land gained from the Mediterranean at the mouth of the Nile.

It is not in many places—perhaps there is no other instance on the face of the globe—where the proofs of early civilization are so marked and so numerous, or where the date of very ancient monuments can be so clearly traced historically, as in Egypt. The seat of the earliest known civilization, remarkable for the peculiarly enduring character of the material used for the construction of public monuments, and for the mode in which history was expressed in picture writing, there is in Egypt every favourable condition for handing down to the most remote time any events once recorded in the ancient fashion. Connected, too, as that country is with the history of the most advanced and influential races of men,—with the Greeks and Romans, who succeeded the Egyptians as the leading people on the earth, and excelled them in cultivation—there has never been a time when the chain of history was broken, nor has there been any reasonable doubt thrown on the interpretation of the picture language since it was first successfully read.

But in addition to these reasons why Egypt should be selected to determine any question in which the antiquity of the human race is concerned,

there are some peculiarities of physical geography
for which the country is also remarkable.   Placed
at the further extremity of the Mediterranean—a
vast inland sea little disturbed by tides, and having
only one small communication with the ocean—
intersected by a river of the first magnitude,
traversing a large continent, and fed by supplies
periodically increasing so as to produce an annual
overflow of the waters, there must have been from
the period when the Nile was first a river, and
Africa a continent, a large deposit of mud brought
down to the mouth of the river, and this in course
of time has become the great Delta as it now
exists.

The Delta is, in fact, the result of this gradual
accumulation of mud brought down and deposited
where the running water of the river first met the
salt water of the sea.   It occupies, or rather it
consists of, the low flat land on the river banks,
still increasing, both in elevation and extension
towards the Mediterranean, at the same slow rate
at which it has always advanced.

There is not much probability that the actual
quantity of mud brought down from the upper
country was ever very different from what it is now;
but at first it was spread over a comparatively
small area, and therefore may have accumulated
more rapidly.   The Delta once originated, the additional height of the land inundated, and of the
bed of the river, is due only to the thin coat of mud

left behind when the waters recede, and must, although very small in each year, become at last appreciable; and thus not only does the Delta itself obtain extension towards the sea, but every part of it gains height. On the running-off of the waters into the Mediterranean, a coat of slimy mud annually covers all the plains, and it is only owing to the rapidity with which the waters evaporate under the hot sun of Egypt, that the land is at all free from miasma and habitable by man; while, owing to the constant accession of fresh soil, it is one of the most fertile tracts on the face of the earth. No doubt such land early attracted settlers, who after a time attained a certain degree of civilization; and from the makers of half-burned brick and pottery, they advanced to be architects, sculpturing the granite from the adjacent rocks of Syene, removing their gigantic blocks of rough, hewn, or sculptured stone to the sites of their towns, and commencing that written history which, during the present century, has been traced by a host of learned philologists, historians, and geographers.

The ancient city of Memphis appears to have been founded many centuries before the time when the pyramids were built, or even when Joseph was first sold into Egypt. It is situated very near the head of the Delta, and where the expansion east and west first begins.

Among the monuments in Memphis (now covered

with the deposits of Nile mud that have been ever
since accumulating) is a remarkable statue thrown
from its pedestal, but lying close to it, and little
injured. The statue is the colossal figure of one of
the Egyptian kings, the second Rameses of
Egyptian history—the Sesostris of the Greeks.
The ruins of Memphis are now partly covered by
a modern village, situated thirty miles above the
point where the triangular form of the Delta first
begins to be determined, and is placed, therefore,
in the true valley of the Nile, between the Libyan
and Arabian hills. The time when Sesostris reigned
in Egypt is believed by Dr. Lepsius to have been
about thirteen hundred and fifty years before Christ,
or about thirty-two centuries ago. The statue is
now buried between eleven and twelve feet below
the surface, and as it is likely that the pedestal of
the statue would have been originally partly below
the surface, we may fairly calculate that in this
particular spot the accumulation of Nile mud has
been not more than eleven feet in the whole period,
being at the rate of a little more than four inches in
each century. As, however, the evidence derived
from a single observation in a matter of this kind
could hardly be depended on, we must seek another
proof at some other point.

The obelisk at Heliopolis, situated close to the
apex of the Delta, is a monument of even greater
age than the statue of Rameses, and is believed to
have been set up more than forty centuries ago.
There seems some reason to suppose that the

temple and city of Heliopolis were built on a site somewhat elevated above the low grounds of the Nile; but, however this may be, the bottom of the pedestal of the obelisk referred to is now twelve feet four inches and a-half below the surface. Estimating the time that has elapsed since the construction of the monument at forty-one centuries and a-half, the average rate of increase would seem to have been four inches per century, but may have been somewhat greater. A further confirmation of this estimate is obtained from the independent investigations of M. Girard, who accompanied the French scientific expedition to Egypt under Buonaparte at the beginning of the present century, and who discovered what was considered to be one of the ancient nilometers of the Egyptians, on which was a mark in Greek characters, bearing the date of the reign of the Roman Emperor SEPTIMUS SEVERUS, who died A.D. 211. The difference of height reached by ·the Nile in its great inundations between that time and the present, appeared to be six feet eleven inches, giving an average of something more than five inches per century. The examination of another nilometer on the island of Rhoda, near Cairo, reconstructed A.D. 847, gives an average of four inches and three-quarters per century.

Although, therefore, it cannot be regarded as a matter about which there is no dispute, all the evidence that exists seems to point to five inches per century as fully representing the average amount of elevation given by the Nile mud to the bed of

the Nile and the surrounding country covered by
the annual inundations.

Mr. Horner, indeed, from various considerations,
sees reason to believe that this amount is in excess,
and that three inches and a-half per century is a
more correct estimate. There is no great probability
that any exact record can be obtained in this way
of the number of years that have elapsed since any
particular part of the mud was deposited, as there
are several sources of possible error; but at any
rate the average can hardly under any calculation
have exceeded five inches per century during the
last several centuries; whilst from the mere effects
of long-continued pressure, the beds must be more
compact at some depth below than they are near
the surface, and the rate of thickness ought to become gradually less the deeper we penetrate.

The object in view in the operations suggested
by Mr. Horner, and carried out under able superintendence, and chiefly at the cost of the Pacha of
Egypt, was to dig a number of pits and bore holes
in the ground through the accumulated mud, at a
sufficient number of points and over a sufficiently
large area to give a fair notion of the nature and
contents of the mud at various places and different
depths, especially where the deposit for some thousands of years must have been tolerably uniform.
For this purpose, seventeen pits were sunk within
the space that may be considered to be the area of the
ancient city of Memphis, whose origin is supposed to
date back nearly six thousand years. Twenty-seven

more such pits were then sunk into the bed of the valley, about one mile broad, east and west of the site of Memphis, between the Libyan hills on the west and the Arabian hills on the east of the Nile. Afterwards the site of the ancient city of Heliopolis, twenty miles below that of Memphis, being selected as a second point, twenty-six pits were sunk on the left or west bank of the river, towards the Libyan hills, and twenty-five more on the right bank, towards Arabia. Thus no less than ninety-five pits in all were sunk, each pit being five feet square, until water came in, and then continued to variable depths up to sixty feet by boring.

In no one instance did the sinking or boring reach solid rock, nothing being sunk through but coarse sand and ordinary Nile mud. We may thus safely regard the delta of the Nile, although one of the most modern events of geological history, as dating back some hundred and fifty centuries at least; and this of itself is one not unimportant result obtained by these operations.

The water was generally reached at less than twenty feet, but the depths were very irregular, one filling at ten feet, and another being dry at thirty feet, although the distance asunder was but short. There appears, indeed, to be no regular depth at which water can be calculated on in any part of the Delta, although the nature of the muddy deposit is remarkably uniform. This arises perhaps partly, if not entirely, from the extreme irregularity of the sands that alternate with the mud.

The depth of ground excavated—the space over which the pits were sunk and the ground bored, and the very large number of sinking and boring operations carried on—give extreme interest to this investigation, and render its results very valuable. Among the most remarkable of these results, and that to which our attention must now be directed, was the large number of cases in which remains showing the handiwork of man were brought up from considerable depths. To illustrate this point, we give an abstract of some of the sinkings.

In a pit sunk close to the statue of Rameses II. the following were the beds sunk through, and their contents :—

	ft.	in.
Brown sandy argillaceous earth, with a mixture of white sand and small fragments of limestone and burnt brick	2	7½
The same with fragments of pottery	6	10½
The same with fragments of limestone and brick . .	2	3½
The same with fragments of burnt brick	4	7¼
Filtration water at 16ft. 4¾in.		
Light brown sandy argillaceous earth, with fragments of burnt brick	0	8
Dark brown argillaceous earth, with fragments of burnt brick and particles of pottery	4	0
The same with fragments of limestone	4	0
The same with fragments of burnt brick and pottery	2	0
The same with fragments of burnt brick, pottery, and limestone	1	0
Argillaceous earth with quartzose sand	2	0
Shining black sand (magnetic iron and quartz) . . .	2	0
The same with argillaceous earth	4	0
The same, but with fragments of burnt brick and pottery	2	0
Sand, with argillaceous earth	2	0
Total depth	40	0¾

Amongst the objects brought up were the following :—At six feet part of a small human figure, and at ten feet part of a small figure of a lion, both in baked clay; at seven feet bones of a jackal, dog, ass, and dromedary; at ten and twelve feet parts of a Nile shell and a marine shell (a *Spondylus*). Pottery was found at various depths, especially at six, eight, ten, fourteen, and fifteen feet; that at fourteen being white, and the rest coarse unglazed pots, jars, or saucers. At eleven feet a fragment of a jar was found with a stamped ornament, at twelve feet a small fragment of coloured mosaic, and at thirteen feet the blade of a metal knife, made of copper hardened with a little arsenic. Fragments of burnt brick and pottery continued through the borings, at intervals, down to thirty-eight feet, below which the search did not go.

There is much that is very curious in the discovery of these remains, which show a considerable degree of art and civilization in deposits which, if the accumulation of Nile mud is a measure of time, must have dated back at least thirty centuries from the present time in the case of the hardened metal knife; twenty-four centuries in that of the little figure of an animal in baked clay; and no less than ninety centuries in that of the fragments of pottery. It is clear that more evidence was required before such conclusions could be at all admitted. The spot selected might have been for some reason the receptacle of fragments of a much

more recent period. There might have been a channel
dug near here for some purpose, and since filled up,
or other possibilities that need not be suggested.
Accordingly, another pit was sunk about two years
afterwards to verify the first, and then seventeen
other pits in various parts of the area of ancient
Memphis. In this second pit the water was removed until the pit was sunk five feet square to
twenty-four feet four inches, and then a boring
was carried down seventeen feet. Here also bones
of domestic animals (ox and hog) were found at
thirteen and twenty feet; the neck of a vessel of
coarse unglazed pottery at twenty-one feet; a
fragment of a small glass vessel and a cube of coarse
sandstone, smoothed, at twenty-four feet; and from
the lowest part of the boring, a fragment, about an
inch square, of unglazed red pottery.

On analysing the earth at the top and bottom of
the excavation, and comparing it with the composition of Nile mud generally, there was found no
essential difference; and thus we are driven to the
conclusion that some three thousand years ago a
statue was erected on a pedestal on about three
feet of sand artificially placed (rendering it probable that the pedestal itself was buried some depth
below the general surface), on a spot where no perceptible change has since occurred to the foundation
or underlying earth, but which has since been
covered year after year by the same Nile mud till
about twelve feet have been accumulated. On both

sides, about eight yards east and west of this pedestal, pits being sunk, show at intervals, to a depth of nearly thirty feet, unmistakeable evidences of human agency, proving an extraordinary amount of civilization. Let us next consider the cumulative evidence obtained from the rest of the ninety-three pits sunk here and in other parts of the Nile valley.

We have first the seventeen pits sunk in various places over the supposed area of ancient Memphis. These range up to three quarters of a mile south, a mile east, a hundred and fifty yards west, and about a third of a mile north of the position of the statue. In all the pits and borings, without exception, the material penetrated was that peculiar argillaceous earth which we call Nile mud, together with occasional sand. The results may be thus briefly enumerated. In a pit twenty yards west and forty-five south of the statue, at thirteen feet down were found sculptured granite and architecturally carved limestone. In another, about twenty yards east and fifty north, at the same depth, a female foot, carved in white limestone, and an ornamental vase of red pottery. In another, eighty yards west and two hundred and sixty north, several successive courses of sandstone to a depth of nearly twelve feet. In another, thirty-five yards east and three hundred and twenty-six north, various statuettes at eight feet and a half to fifteen feet, and at thirty-three feet and a half, a tablet with inscriptions and fragments of pottery. With

the exception of the tablet in the last pit referred
to, nothing can show more clearly that at a depth of
from eight to twelve or fifteen feet from the present
surface, and occasionally a little below, are the
remains of the ancient city of Memphis, and if, as
is supposed, this city was founded nearly three
thousand years before the erection of the statue
now twelve feet below the surface, we should have
a ready explanation of any discoveries to a depth
of twenty-four feet, and the chance of occasionally
or accidentally buried fragments to a greater depth.
Out of three others of the seventeen pits carried
down much below the level of the base of the pedestal,
fragments of pottery were taken, but none of these,
and none of the more shallow pits, yielded sculptured
stones below a very small depth.

Passing on now to the more distant pits we
obtain evidence of a different kind. Commenced at
different levels, they were none of them sunk twenty
feet from the surface, and some were much less;
all were sunk through the same argillaceous earth
and occasional quartzose sand, fifteen of them being
either almost or entirely in that clayey sediment
recognised as Nile mud, and eleven chiefly in
quartzose sand.   Fragments of burnt brick were
found at the bottom of a pit sunk thirteen feet
through Nile mud, but with one or two exceptions
there was no additional proof of human ingenuity
having been exerted in the construction of works
of art. It would seem, therefore, that in the actual

vicinity of the ancient city of Memphis the remains
are very abundant, while at a distance to the right
and left they are far more rare.

At a distance of about five miles and a half
N.N.E. of Cairo, and less than four miles from the
right bank of the Nile, is the obelisk of Heliopolis.
This most ancient monument is supposed to have
been erected about 2300 years before Christ, and it
marks the site of the city of " On," whose walls
are still traceable as mounds of clay formed originally of unburnt brick.   Eastward of this obelisk
the ground rises abruptly, and is out of the reach of
inundation, and it seems likely that the site of the
city was chosen as being at the time also untouched
by the waters of the Nile at its highest.   An excavation close to the obelisk shows as follows :—

	ft.	in.
Disturbed ground mixed with rubbish	0	9
Undisturbed Nile sediment, with occasional fragments of pottery	4	11
Rubbish soil (blackish brown earth with fragments of limestone)	4	10
Coarse grey sand (with bones in its upper part) . . .	12	2

Of this depth about ten feet six inches may be
regarded as Nile mud, and in adjacent pits the
thickness varied, becoming greater towards the
west.   In one pit and boring, close to the banks
of the Nile, the borings reached to fifty feet, and
in one instance to sixty feet.   In the latter case
more than half the boring was through sands, probably blown in from the desert, and quite indepen-

dent of the annual Nile accumulation; but there still remains thirty feet of real mud, and the rock underlying the Delta was not reached. Fragments of pottery appear to have been found at the depth of nearly fifty feet in one of three deep pits, and at the lowest point at which Nile mud was reached in another, and fragments of burnt brick accompany the pottery at a somewhat smaller depth. No doubt the accumulation may have been more rapid close to the raised bed of the Nile than on the lower land between the banks of the river and the hills; but still the thickness of mud cannot but represent an enormous lapse of time, during the whole of which the human race must have inhabited Egypt.

In the course of this large and important series of excavations, all made by intelligent persons aware of the object of investigation, but in no way likely to misrepresent facts, the discoveries were very varied and important, and all seem to tend to one conclusion, the great age of the deposit. The extremely uniform accumulation of the sediment, the occasional alternation of loose sands with the mud, these sands not mixing with the sediment, but forming distinct strata; the total absence anywhere of solid rock, or any fragments of any kind that would show the presence of a disturbing cause during the formation of the Delta; the extreme rarity of organic remains, except immediately under the ancient cities, and then at no great

depth; the total absence of any extinct species of
animal of any class; and lastly, the most remarkable fact of all, the discovery at all depths and in
almost all parts of the ground experimented on of
remains of burnt brick and pottery indicating
beyond a shadow of doubt the residence of civilized
races of men:—These are all indications which
cannot be neglected, and taken in connexion with
what is known historically of Egypt, and of the
formation of the Nile Delta, they throw back the
history of civilization to a very distant epoch.*

It is well to remind the reader that the speculative dates assigned to the ancient cities of Memphis
and Heliopolis by the German authorities quoted,
although correct according to the most probable
interpretation of documents and hieroglyphics,
really hardly affect the question here involved as to
the great antiquity of the human race in Egypt.
There is not, and never was, a shadow of doubt that
the two monuments referred to are among the very
ancient decorations of the long extinct cities of
Memphis and Heliopolis, which, even in the time
of Herodotus, were objects of antiquarian investigation. The existence of undisturbed Nile sediment

* The accumulation of material that appears always to take
place on the site of a large town, and which is due to the materials transported thither, and left by the successive inhabitants,
may no doubt account for some part of the thickness of earth
around these ancient cities. This it is not easy to calculate; but
it could hardly affect to any considerable extent the land completely across the Nile valley, over the whole of which similar
human remains occur.

below the foundations of these cities, of much greater thickness than the mud that has since accumulated, and now covers them, and the presence in this undisturbed mud of human remains irregularly distributed over a large area, affords distinct proof of the existence of civilized races at a period long antecedent. Long before this, again, and for centuries, if not thousands of years, there must have been uncivilized tribes utterly unacquainted with the arts of pottery, or the habits which could enable them to make use of such arts. There is no conceivable method by which the true Nile mud, remarkable in its composition, and not to be mistaken for other kinds of mud, could be accumulated regularly, as we find it throughout the valley of the Nile near Cairo, except by the annual inundation, and the layer deposited each year by the inundation is so thin, so uniform, and so regular, that it does not admit of much change.

Without, then, venturing on a calculation of the actual number of years, or centuries, we are thrown upon the conclusion that a very much longer period is required for the duration of the human race than is allowed by the ordinary chronology, if we attempt to satisfy the conditions of recent discovery in Egypt.*

It ought not to be necessary to remark that such

---

* Although the present volume hardly seems the place to discuss another very important argument in favour of the great

a discovery, however it may shake our faith in the recognised chronology of the Bible, does not in the smallest degree affect one particle of the evidence according to which we believe the Bible to be a sacred book, or rather a collection of sacred records handed down to us as the sources of all moral and religious truth. The history of the human race, as given us in that book, is far too slightly sketched, and the truths contained in it are too independent of such matters to justify any alarm. Just as in astronomy and geology, and even in general history, there is found in the Bible the current language of the time when the various books were first collected into a roll in the latter part of Jewish history,

antiquity (compared with the ordinary chronology) of the human race, the following extract from the preface to the third (the last published) volume of Bunsen's elaborate work already referred to (" Egypt's Place in Universal History") cannot fail to interest the reader, not only as supporting the conclusion arrived at, but for its own sake. After mentioning the results of Mr. Horner's investigations, the author proceeds to remark "that historical facts lead to the same conclusion, if the space of time during which man had existed on our mother earth be measured, not by conventional notions arising out of ignorance and sanctioned by prejudice, but by facts which any one is capable of investigating who does not shrink from researches, determinable with logical demonstration and mathematical cogency. The indisputable facts of the development of language suffice to prove the two points at issue, that the period commonly assigned to the existence of mankind is much too brief, and that the real duration is not immeasurably or indefinitely long. The author would speak freely on this subject, because he feels strongly that in the times in which we live it is as absurd and as irreverent to ignore the linguistic strata as it would be to take no notice of the strata of the earth, or for a man to set up a system of astronomy of his own, without reference to the Keplerian laws, or Newton's immortal discoveries. Much certainly remains to be done before so the chronology, as far as any clear allusion to actual time is to be traced, is undoubtedly that of the same period. No miraculous information— nothing beyond the current ideas and language of the day—is in any place vouchsafed with reference to those points of investigation for which human inquiry is sufficient. Astronomy is not corrected there, geology is not alluded to, and chronology is equally left to be determined by the aid of those faculties with which we are endowed.

---

the two kinds of research bearing upon this point of chronology are completed and consolidated."

As regards the historical inquiry, the author also states that his investigations have led to results identical with those flowing from Mr. Horner's researches. "They are based principally on the history of the language of Asia, and their connexion with that of Egypt, and they do not in his opinion contravene in the slightest degree the statements of Scripture, though they de-molish ancient and modern rabbinical assumption, while, on the contrary, they extend the antiquity of the biblical accounts, and explain for the first time their historical truth."—*Egypt's Place in Universal History*, English translation, vol. iii. p. xxvi.

XI.

HUMAN REMAINS IN CAVERNS AND GRAVEL.

Nature of limestone-rocks—Origin of the caverns with which they abound—Dr. Buckland and the caverns of Devonshire and Yorkshire—Contents of Kirkdale cavern—Opening of the Brixham cavern in 1858—Discovery of remains of human art with extinct bones—Description of the objects found—Mode of construction—Discovery of similar implements, or weapons, in Suffolk, in 1797, and in France in 1840—Investigation by Mr. Prestwich and Dr. Falconer—Additional evidence adduced—Objections answered.

LIMESTONE rocks formed in water, as they all appear to have been, and brought into their present condition by a long process of partial drying and constant upheaval, being for the most part brittle rocks, abounding in crevices and fissures, and particularly subject, both to the dissolving power and mechanical action of water, generally abound in cavities and open spaces, except where the rock is too soft to serve as a roof to large spaces. Thus chalk rarely contains them, and the softer kinds of limestone only occasionally, and of moderate size, while the harder limestones are remarkable for the number, magnitude, and varied form of their caverns, sometimes extending for miles, with deep holes full of water, and often the scene of fairylike and fantastic columns and sheets of limestone, the result of the partial evaporation of water charged with carbonate of lime in solution, which has trickled through the fissures into the cavity. The great Mammoth Cave, in the carboniferous limestone of Kentucky, the extraordinary and extensive caverns of Adelsberg, in the half-crystalline Alpine limestone of Carinthia, the cavern or grotto of Antiparos, and the labyrinth of Crete in Greece, are all well-known and magnificent examples of the largest class. The caverns of Franconia, in the oolitic rocks of Franconia, in Germany; those of the carboniferous limestone of Derbyshire, Yorkshire, and Devonshire, in England, and those of Central France, celebrated for the ripening of Roquefort cheese, are almost equally well known in the countries in which they occur.

It is now nearly forty years since Dr. Buckland, at that time commencing his career as a geologist, undertook the examination of some remarkable caverns in Devonshire and Yorkshire, in which the bones of a large number of animals had recently been discovered. Cuvier had not long before described the marvellous contents of the quarries around Paris, and had made out the fact that the quadrupeds whose remains are there so abundant belonged to species and genera exceedingly unlike the existing races. The cavern animals, however,

were far more modern, and though the bones indicated extinct species, and even genera now absent
from the country, they were found to depart far less
from the existing forms, as they included bears,
hyænas, and the carnivorous races, with varieties
of deer and some of the smaller quadrupeds. There
were also bones of the elephant, rhinoceros, and
hippopotamus. Amongst these, not so much in
England as in the South of France and in Germany,
it was stated from time to time that there were
indications of human remains, not indeed bones,
but weapons which rude uncivilized tribes might
have used. The following paragraph, extracted
from Dr. Buckland's book, will show the state in
which the bones of animals were found. It refers
to one of the most remarkable of the Yorkshire
caverns, that of Kirkdale, then recently opened,
and the description will apply to the similar caves
elsewhere, especially in England and Belgium,
where the phenomena are very uniform.

" The bottom of the cave, on first removing the
mud, was found to be strewed all over, like a dog
kennel, from one end to the other, with hundreds
of teeth and bones, or rather broken and splintered
fragments of bones, of all the animals above enumerated. They were found in greatest quantity near
its mouth, simply because its area in that part was
most capacious. Those of the larger animals, elephant, rhinoceros, &c., were found co-extensively
with all the rest, even in the inmost and smaller

recesses. Scarcely a bone has escaped fracture. On some of the bones marks may be traced, which, on applying one to the other, appear exactly to fit the form of the canine teeth of the hyæna that occur in the cave."

"The jaw-bones are broken to pieces like the rest, and in the case of all the animals, the number of teeth and of small bones of the extremities, is more than twenty times as great as could have been supplied by the individuals whose other bones we find mixed with them."

"The greatest number of teeth are those of hyænas and the ruminantia. Mr. Gibson alone collected more than three hundred canine teeth of the hyæna, which must have belonged to at least seventy-five individuals, and adding these to the teeth I have seen in other collections, I cannot calculate the total number of hyænas of which there is evidence at less than two hundred or three hundred." After mentioning further details, Dr. Buckland states his opinion to be "that the cave at Kirkdale was, during a long succession of years, inhabited as a den by hyænas, and that they dragged into its recesses the other animal bodies whose remains are found mixed indiscriminately with their own," this conclusion being confirmed by the discovery of the solid calcareous excrement of some animal that had lived on bones, of which considerable quantities were also met with, either detached or invested with a crust of stalagmite,

and which was recognised by the keeper of a mena-
gerie as identical with the excrement of recent
hyænas. He concludes that "the accumulation
of these (the hyæna and other) bones appears to have
been a long process, going on through a succession
of years, whilst all the animals in question were
natives of this country." "The teeth and fragments of bone seem to have lain a long time
scattered irregularly over the bottom of the den,
and to have been continually accumulating, until
the introduction of the sediment in which they are
now imbedded, and to the protection of which they
owe that high state of preservation they possess."

Finally, the professor considers that four periods
of time are indicated by the condition of remains
in this cave. "First, when the cavern and its
opening existed in its present state, but was not
tenanted by hyænas. This is considered to have
been very short. Second, when the cave was inhabited by hyænas, and the stalactite and stalagmite
were still forming. Third, when the mud was introduced and the animals extirpated; and fourth,
when the stalagmite was deposited which invests
the upper surface of the mud."*

At the time when these accounts were published,
it was considered in England that no evidence had
been adduced in favour of a view, to some extent
current on the Continent, that human remains

---

* Buckland's *Reliquiæ Diluvianæ*, pp. 15—54.

occasionally accompanied the cavern bones. The caverns in some of the limestone districts of France had indeed seemed to show that remains of human art were present there, and belonged to the period of these hyænas and bears, but the subject was never fairly met, and the evidence was lost. From the time of the publication of Dr. Buckland's book till the year 1858, the caverns and their contents yielded little of importance to geologists, and that little did not touch the subject of human remains: but at that time there was a startling announcement in reference to a cavern at Brixham, in Devonshire, one of a numerous class often examined, but which was thought worthy of careful and minute investigation. Aided by a grant from the Royal Society, and under the superintendence of Mr. Bristow, of the Geological Survey, and Mr. Pengelly, a well-known cavern searcher—Dr. Falconer also giving advice from a distance and occasionally on the spot—this cavern was cautiously and conscientiously opened. It is nothing more than a crevice of no large size, one of a class common in the carboniferous limestone, which is a rock divided into joints crossing each other at right angles, and, as these are especially subject to the percolation of rain-water, the stone is often eaten away into hollows. The action of the rain-water may have been assisted by streams, and even by the sea, when the level of the land was lower than at present. The floor of the cave is a coarse pebbly deposit, over which is a loamy deposit, partially covered with stalagmite and partially overlaid by other material. Among the loam, and sometimes under the stalagmite, are numerous remains of extinct animals, common in the caverns of Somersetshire and Yorkshire, and with these, evidently deposited at the same time, in the middle of the cavern, under stalagmite itself, and actually entangled with an antler of a reindeer and the bones of the great cavern bear, were found rude sculptured flints, such as are known to have been used by savages in most parts of the world, as lance and arrow heads. Although of a rude form, these are so unmistakeably artificial, that no doubt can be felt on the subject.

At the very time that these investigations were going on, Dr. Falconer was obliged to leave England for the winter, and in Sicily visited a cave previously undescribed. There he discovered, on the roof of the cavern, which was nearly empty, a large patch of bone breccia (a mass made up of fragments of bone cemented by stalagmitic limestone), containing teeth of ruminants, bits of carbon, shells of various species of snail, together with a vast abundance of flint and agate knives of human manufacture. Of these latter, the great majority present definite forms, being long, narrow, and thin, having invariably a smooth conchoidal fracture below and a longitudinal ridge above, bevelled off right and left, or a concave facet replacing the ridge; in the latter case presenting

three facets on the upper side.   They are considered
by Dr. Falconer to resemble closely, in every detail
of form, obsidian knives from Mexico, and flat
knives from Stonehenge, Arabia, and elsewhere,
and they appear to have been formed by splitting
off long angles of prismatic blocks of stone.   These
fragments are intimately intermixed with bone-
splinters, shells, &c., and the bones are those of
animals usually found in caverns, having belonged
to wild carnivora living in the caverns as dens,
or else to those ruminants and other animals
serving as the prey of these carnivora.   It is clear
that the original deposit, containing the admixture
of flint knives with bones and shells, must have been
drifted in, at a very ancient period, in tranquil water
—that the cavern must have been completely filled
with this material—that stalagmite had formed
amongst it—that it had subsequently been swept
out to a great extent—and since then, that the
whole rock has been greatly altered in position.

In several caves in the South of France, and in
some of those in Franconia (Germany), there have
been described, from time to time, collections of
fragments, apparently indicating a race of man very
little civilized, and antecedent to those of which
there are historic records.   Each has been in
turn explained away; and, as a general impression
prevailed that the matter was beyond discussion, a
very little doubt was enough to cast discredit on
any report.   Now, however, that bone caverns con-

taining sculptured flints, with similar human remains, have been discovered not only in England,
but in Italy, and that these have yielded to careful
observers such human remains mixed unmistakeably
with bones of extinct species of animals, in such a
way as to show that all were originally drifted and
buried together, the former cases require reconsideration. Unluckily, the evidence is lost.

At the time when the caverns are supposed to
have been occupied by the extinct races above
alluded to—the hyæna, the bear, and the tiger; the
mammoth, or early elephant, the tichorhine rhinoceros, and the hippopotamus; the Irish elk, and
extinct horses and horned cattle—there was a large
and important formation going on, chiefly consisting
of rolled and water-worn fragments of flint, mixed
with occasional angular blocks of hard stone and
sand. In this formation, the result of numerous
glaciers proceeding from adjacent hills and shores,
and icebergs floated off from adjacent coasts,
and caught on submarine banks as they journeyed
southward loaded with *débris*, are found also bones
and other indications of ancient tribes of animals
now altogether passed away, and with all these are
indications of human agency precisely of the same
kind as has been already alluded to—namely, arrow-
heads, lance-heads, knives, and other weapons and
utensils constructed of flint.

The precise nature of the implements thus rudely
constructed of flint, which are assumed to mark the

existence of a tribe of men contemporaneous with the extinct quadrupeds, and living in or near the place where we now find them, is a matter of considerable interest, and one on which there has been some discussion, both geological and archæological. The very essential fact of their artificial construction—of their really being broken intentionally by the hand of man for a special purpose—has been questioned; and no doubt the indications of contrivance in a number of the specimens collected are very faint. Their object and use have been the subject of much discussion, and theories have been based on some fancied resemblance of some of them to objects of more modern construction. The use to which they were applied, if really of human manufacture, is, however, a matter of no importance to the geologist; while the mere existence of a single one out of the whole number which can be shown to have been deposited with the bones is sufficient to prove the geological assumption, and at least renders probable a similar origin for more questionable specimens.

The fractured stones attributed to human agency, hitherto discovered in caverns and gravel, include a large number of flakes or chips, chiefly of flint but occasionally of granite, porphyry, jade, serpentine, jasper, basalt, and other hard tough kinds of stone; a smaller number of more artistically cut specimens, supposed to have been arrow-heads, spear or lance-heads, knives or hatchets of similar stone;

and some few, yet more carefully finished and comparatively perfect, specimens of the same kind. These curious fragments of ancient art are described as referrible to the three following forms—namely; 1. Flakes of flint, apparently intended for knives or arrow-heads. 2. Pointed implements, usually truncated at the base, and varying in length from four to nine inches, possibly used as spear or lance heads, which in shape they resemble. 3. Oval or almond-shaped implements, from two to nine inches in length, and with a cutting edge all round. They have generally one end more sharply curved than the other, and occasionally even pointed, and may be supposed to have been used as sling stones, or as axes, cutting at either end, with a handle bound round the centre. This description of the objects in question was given by Mr. Evans, at a meeting of the Society of Anti-quaries, in June last, and is the more to be valued, as Mr. Evans accompanied Mr. Prestwich on his visit to the neighbourhood of Abbeville, to examine some remarkable quarries there which have yielded numerous implements mixed with bones of extinct quadrupeds. Mr. Evans further states, "that the evidence derived from the implements of the first form is one of much weight, on account of the extreme simplicity of the implements, which at times renders it difficult to determine whether they were produced by art or by natural causes. This simplicity of form would also prevent the flint

flakes made at the earliest period from being distinguishable from those of later date. The case is
different with the other two forms of implements,
which were unquestionably worked by the hand of
man, and are not indebted for their shape to any
natural configuration or peculiar feature of the
flint." They present little analogy in form to the
well-known implements of the so-called Celtic or
stone period, which, indeed, are generally smoothed,
or even polished, and are made of various kinds of
hard stone. Those from the drift and caverns are
never smoothed, and have not yet been found of
other material than chalk flint.*

M. Boucher de Perthes, the first discoverer of
these, whose statement deserves every consideration,
since all that he has professed to describe appears
to have been confirmed by later observers, is satisfied
that he has found in many cases (of course greatly
injured by time) the handles of wood and stag's-horn
originally attached to some of these implements.
He believes that these were of the simplest kind,
and he supposes that they were actually buried
with the flinty and more enduring part, and are
only not now exhumed owing to their more perish-
able material. However this may be, the negative

---

* Mr. Evans further states that in form and workmanship the
flint implements discovered at St. Acheul (Abbeville), differed
essentially from those of the so-called Celtic period, and that had
they been found under any circumstances, they must have been
regarded as the work of some other race than the Celts, or
known aboriginal tribes.—Proc. Roy. Soc. for May 26th, 1859.

fact of no other remains occurring cannot fairly be
held to militate against the evidence afforded by
their presence, and the specimens show every appearance of having been fabricated by another race
of men than the Celts. Judging from the fact that
Celtic stone weapons of more finished make, and
mixed with pottery, have been found in the superficial soil above the drift, or in beds separated by distinct deposits of stalagmite from those containing
these ruder weapons, the tribes who constructed and
buried the latter must have inhabited this region
of the globe at a period anterior to its so-called
Celtic occupation.

It is well worthy of notice, that in all parts of
the world in which archæological researches have
hitherto been made, implements closely resembling
either those above described, or the more finished
and smoother weapons of a similar character, but
later date, referred in England to the Celts, have
been found buried, or are still used by tribes little
advanced in the arts of civilized life.

Thus the flint axes and knives of the caverns or
the gravel scarcely differ in form from the hatchets
of the latest inhabitants of Britain previous to the
incursion of the Saxon and Danish tribes; but the
former are exclusively flint, and the latter show a
large admixture of other stones, many of which are
foreign. The arrow-heads and hatchets of England differ in no respect whatever from those of
Normandy and Brittany, of the centre and south

of France, of Spain and Italy, of Sardinia and
Sicily, and of other islands, as well as many places
on the shores of the Mediterranean. These again
have their exact counterparts in Arabia, India, and
Eastern Asia, in the islands of the Indian Archipelago and the South Pacific Ocean, in Mexico and
the West-Indian Islands, and on the mainland of
America. Stone hatchets from Virginia are precisely identical in form with the Celts of England,
and in every case these, and a few straggling fragments of the coarsest pottery, form the principal
objects that remain to illustrate the manufactures
of races of people essentially different, inhabiting
the most widely distant countries for many ages.

The method that seems to have been adopted in
the manufacture of the older flint and stone implements was simple enough, and can be followed now
with success by any one possessed of sufficient time
and patience. The stone was probably taken from
the rock, as in that case it would be rather softer
than after long exposure to the air. By a number
of slight blows, made by using one stone as a
chisel, and another as a hammer, small chips were
knocked off in the right direction, and thus the
number of the faces becomes a proof of the mode,
as well as of the fact, of manufacture. In after
times the edges were rubbed down on another
stone, as hard as or harder than the implement to be
constructed; and at a later period still, a rough
polishing process was introduced.

These latter steps were, however, in all probability, modern innovations, and involved efforts far
beyond the powers of the older tribes. Not only
are the implements and weapons found in various
stages of completeness, but very rough beginnings
are sometimes seen, and whole baskets-full of chips
have been described as occurring in some localities.
All of these tell the same tale, and although no
doubt each one possesses a special interest, little is
gained to science by the repetition of specimens,
either alike in structure and use, or intended for
purposes we cannot at all make out. The main
fact is this,—that flint and other hard stones bearing marks of having been broken artificially into a
definite form, afford proof of the existence of human
beings at the time when the stones were thus
wrought into shape.

There is one other point of considerable importance to be referred to before leaving this part of
the subject. It is the nature of the stone itself.
In many of the later and more perfect specimens of
manufacture, there is some difficulty in tracing the
stone to any neighbouring quarry, and not unfrequently the material is decidedly and unmistakeably
foreign. In all the oldest flakes, arrow-heads, knives,
and hatchets, the stone is that of the vicinity; flint
in England, where this mineral is common, basalt
or granite where flint is not obtainable. Elsewhere we find other rocks, but always the hardest
and toughest that could be readily obtained.

The condition of the flint implements in the gravel and caverns is peculiar, and corresponds in weathering with that of fragments of shapeless flint buried with them. They are discoloured by contact with ochreous matter, whitened when in a clayey matrix, incrusted with carbonate of lime when found with chalky matter, just as the other flints are, and in all respects they appear to have undergone the same alteration by time and exposure.

There is naturally a great disinclination on the part of many to accept, on any terms, facts so adverse to all ordinary notions of human chronology; for, as we shall see presently, there is but one conclusion to be arrived at if the artificial origin of these implements is admitted. The races of men that constructed them must, in that case, have been cotemporaries with the great extinct cavern bears, hyænas, and tigers spread over northern Europe at a period when elephants, rhinoceroses, and hippopotamuses wandered over the forests and tenanted the rivers.

Mr. Wright, no mean authority in antiquarian matters, has endeavoured to throw doubt on the artificial origin of the flakes, or more simple implements, of which the number found is very large, and states his belief "that they might have been produced naturally by a violent and continued gyratory motion, perhaps in water, in which they were liable to be struck by other bodies in the same movement." This is a somewhat vague hypothesis, not supported by experiment or observation; but whatever its value may be as regards them, the more artificial forms already found, however few in comparison, are amply sufficient to destroy the value of this doubt, as affecting the general question. With regard to the more highly finished of those described by Mr. Evans, they show a uniformity of shape, a correctness of outline, and a sharpness about the cutting points and edges, which, in the opinion of that gentleman, and most of those who have examined them, could not possibly be the result of accidental collision with other flints. The progress of discovery cannot fail very soon to settle all doubt in this matter, and a very few additional discoveries will decide whether the specimens supposed to be artificial were really the work of intelligent beings, however low in the scale of civilization.

Mr. Wright has also objected that the mere number of the flakes found in the same locality renders it impossible that they can be other than natural phenomena. That they are unequally distributed is, however, very probable, and it is not difficult to understand that objects of this kind might be drifted into certain localities rather than others, whether accumulated by burial, by the accident of local manufacture, owing to an abundance of the raw material, or merely brought together by the separating action of water grouping together objects of similar dimensions and specific gravity. However this may be, the whole question as to the origin of these flakes, knives, and axes must, as we have just observed, be decided shortly by further discoveries.

It was at Abbeville, and as long ago as 1840, that M. Boucher de Perthes, whose labours have been already alluded to, first made the discovery that these supposed human remains existed in the gravel of Abbeville.   Some forty years before that, however (in the year 1797), there had actually been published in England an account, by Mr. John Frere, of similar objects in the gravel of Hoxne, in Suffolk.*   No attention was excited by this notice; nor indeed was the elaborate octavo volume, loaded with illustrations, published in 1847 by M. Boucher, at all more fortunate.   That clever and persevering author, however, feeling that he had right on his side, did not relax in his endeavours to throw fresh light on the important question involved.   In 1851 a part of M. Boucher's work was translated and

---

* The weapons discovered by Mr. Frere, and figured by him in the "Archæologia," were found in a gravel bed two feet thick and twelve feet from the surface.   Above the gravel was a sand, one foot thick, containing marine shells and bones of gigantic land animals, and above this sand was seven feet six inches of brick clay, for the purpose of getting which the ground was opened.   The clay was covered with eighteen inches of vegetable soil.   The weapons were in great abundance—five or six in the space of a square yard, and were mixed with fragments of wood, which decayed on exposure.   Mr. Frere remarks that "the situation in which they were found may tempt us to refer them to a very remote period indeed, *even beyond that of the present world.*"—*Archæologia*, vol. xiii. p. 204.

published in England, and in 1854 Dr. Rigollot, a French geologist, entered fairly into the subject, and satisfied himself as to the geological age of the deposit of gravel, in which no one doubted the remains had been found. The result of his investigations is expressed in a few words in a letter to M. Boucher, dated 29th November, 1854. Speaking of a memoir he is preparing, he says :—" In this memoir, I merely follow in your steps, and my only ambition is to prove that you were correct in announcing that our country had been inhabited by men before the grand disturbance that caused the destruction of the elephants and rhinoceroses that lived there. What you have said, with all the detail required to produce conviction, I have repeated more briefly, and no doubt less well."

M. Rigollot was soon followed by others, and in 1857 a second volume was published by M. Boucher, with fresh evidence, and new figures of sculptured flints discovered, extending also the district containing them, which then included the department of the Somme, the Pas de Calais, the Oise, the Seine, and the Seine Inferieure.

Meanwhile, as we have said, our own geologists were beginning to have their attention directed to the subject. The cave discoveries already alluded to had paved the way for the reconsideration of the evidence, and Dr. Falconer, satisfied that M. Boucher de Perthes deserved some credit for his investigations and specimens, warmly engaged Mr.

Prestwich to examine the Abbeville sections. Mr. Prestwich, in a memoir read 26th of May last, before the Royal Society, confesses that he undertook the inquiry full of doubt, but went to see and judge for himself. He found the gravel beds of St. Acheul (those that had been most productive in flint implements) capping a low chalk hill a mile S.E. of Amiens, above one hundred feet above the level of the Somme, and not commanded by any higher ground. The upper bed consisted of about ten to fifteen feet of brown brick earth, containing many old tombs and some coins, but without organic remains. Under this was a whitish marl and sand, with recent shells and mammalian bones and teeth, whose thickness varied from two to eight feet ; while lastly, there was found six to twelve feet of coarse sub-angular flint gravel, with some remains of shells in sand, and teeth and bones of elephant, horse, ox, and deer, generally near the base. With these he found the worked flints in considerable number, and the whole deposit rests on chalk.

Another section of greater interest is described by Mr. Prestwich in the same memoir. It is at Menchecourt, a suburb to the north-west of Abbeville, where the deposit is very distinct in its character, and the association of flint implements with recent shells and extinct mammalian remains is unquestionable. The mammalian remains here include two extinct deers, an extinct species of

horse, and another of *Bos*, besides the mammoth
and tichorhine rhinoceros, and there are a few marine
shells mixed indiscriminately with fresh-water
species. Amongst these, in the middle beds, at
depths varying from sixteen to twenty-two feet
from the surface, are flint flakes of doubtful character, flint knives resembling those found in barrows,
and recognised by archæologists as of artificial
make, and true flint implements ("haches"), believed by M. Boucher de Perthes to be from the
lower bed of sub-angular flint gravel containing
the mammalian remains.

With regard to the probability of these implements having been accidentally buried in the gravel
since the time of its formation, it is an important
fact that they are stained and coloured like the
other flints in the gravel containing them.

All observers are agreed as to the geological age
of the gravel itself, and of the drift deposits; and
English geologists identify the beds of Abbeville
with those of East Croydon, Wandsworth Common,
and other places near London. M. Buteux, author
of a careful memoir on the geology of the department of the Somme, and M. Hebert, whose special
study has been the deposits of the latest tertiary
period (*terrains quaternaires* of French geologists),
being called on by Dr. Rigollot to examine
rigorously the position of the beds, state that "the
implements are found neither in the loam nor brick
earth forming the upper bed, nor in the intermediate

beds of clay, sand, and small flints, but *exclusively in the true diluvium*—that is, in the deposit which contains the remains of species belonging to the epoch immediately preceding the cataclysm by which they were destroyed. *There cannot be the smallest doubt as to this point.*"*

It is curious to remark with how much difficulty evidence makes its way against preconceived notions. One-tenth part of the testimony that has been produced in reference to the facts just narrated, would have sufficed to admit almost any statement in general science, but even more recently we find another French geologist adding his confirmation, as if the subject were still under dispute. M. Gaudry, a member of the French Institute, accompanied by M. Garnier, librarian of the city of Amiens, and two other gentlemen, went over the same ground as had been previously travelled by Mr. Prestwich and M. Buteux, and the former reported, at a meeting of the Academy on the 3rd of October last, that he caused the face of the quarry at St. Acheul to be opened for the length of seven metres, he himself watching the whole operation, and not leaving the ground while the work was going on. The head of brick-earth, amounting to about one and a half metre, had been removed, and there remained a thickness of two metres before reaching the true drift deposit.

---

* Antiquités Celtiques, tome deuxième, p. 9.

Nothing whatever was found in this overlying bed, but no less than nine of the flint implements were obtained in a flinty bed reposing on white sand, about a metre below the top of the underlying deposit of drift. The flints thus obtained could not have been rolled, their edges being still sharp, and in the same bed, at a little distance, were found remains of rhinoceros, hippopotamus, and mammoth.

Thus, then, there would appear no reasonable doubt that the discoveries of M. Boucher de Perthes, announced some twelve years ago, put prominently before the world in a book written for the purpose of attracting attention to the subject, and supported by specimens offered for inspection to all who would inquire, illustrated also by a vast series of drawings, but quietly set aside by everybody except a few friends of the author, were really among the most remarkable discoveries yet made in geological and archæological science, and were calculated to throw the greatest light on the last great revolutions of the globe. Perhaps, indeed, if M. Boucher had been contented to give his facts simply, or with their necessary inferences, without overlaying them with elaborate explanations, and making them the foundation of theories, he might have been more successful. But however this may be, the facts have at last made their way, and are fairly afloat for general discussion.

To statements of this kind there are always, and

properly, many objections raised. It has been asked why, if the flints are *in situ*, and were certainly manufactured, leading to the inference of human inhabitants contemporaneous with the cavern and gravel animals, are there not also human bones mixed with those of the quadrupeds? To this a very pertinent reply is given in the introduction to M. Boucher's second volume, and it is one which every geologist at least must appreciate. He says—

" Have patience. Before the time of Cuvier, who could have imagined that at Montmartre were hidden thousands of quadrupeds of the older tertiary period? Had any one asserted their existence, and especially the fact that they represented species of animals long since extinct, would not every one have refused belief? And even now would not the very possibility of the event be denied if any one were to assert the recent discovery of a heap of human bones under similar condition? But there is no reason for this, for if not true to-day it may be to-morrow, and if not in Paris or France, it may be elsewhere. Yes, this discovery must take place, and nothing but some retreat of the waters of a lake or a bay, some uplifting of a mountain mass, is needed to produce, not one skeleton only, but thousands; for the abundance of their monuments in stone, their knives and other implements, sufficiently prove that there was a large as well as an ancient population."

XII.

GRAVEL OF THE DRIFT PERIOD, AND ITS CONTENTS.

Meaning of the term Gravel—*Its geological age—Drift beds—Origin of such deposits— Time required to produce them—Tabular statement of rocks of the drift period—Composition of gravel of the South-East of England—Irregular position of drift beds in patches—Agency of ice in transporting the drift—Climate of the period—Distribution of land—Kind of animals existing in Europe—Bats—Carnivorous quadrupeds—Herbivorous quadrupeds—Pachydermatous animals—Ruminants—Corresponding animals of South America and of Australia—Absence of any change in the lower groups—Position of archæology in reference to geology.*

Much has been said in the last chapter, and much has lately been written and talked elsewhere, about the *gravel* of various countries to which Englishmen resort. The term is applied often to very different material, but there is one prevailing idea—namely, that it is that mixture of sand with stones of moderate size, angular or rounded, which is useful for making garden-paths, the foundation of roads, and some other economic purposes. Among geologists the word is used in a somewhat narrower but also in a more inclusive sense—narrower, inasmuch as only those kinds of true gravel-like deposits are

Q 2

comprised which are believed to have been formed during one period, or definite part of the earth's history; but more inclusive, because the admixture of sand with stones, the limited size of the stones and other matters greatly affecting the economic uses of the material, are not considered essential. It is better therefore, in speaking of a particular kind of gravel, to use another term which more strictly marks the date of the formation and leaves the mineral character open. The drift period—the glacial period—the boulder formation, are all terms of this kind. Gravels belong to the whole of that great newer division of rocks called tertiary, and may be found some day in older rocks than any of these. The drift period, which includes all those gravels of which we propose to treat, is not only a part of the tertiary period, but a very modern part.

Speaking technically, it is to the *uppermost plio-cene* or *pleistocene* beds of some British geologists, and the *post-pliocene* or *Recent deposits* of others, that those portions of the gravel containing human remains have hitherto been limited. These beds are of the same age as the glacial drift, or boulder formation of Norfolk and the Clyde valley, and may be conveniently regarded as forming a part, but probably the upper part, of that great deposit of sands, gravels, boulders, and clay marls very widely spread over England and Northern Europe, and designated *drift deposits*. Similar beds occur in the Alps and in the Northern part of North America; beds of the same age, and not very dissimilar in appearance, cover the vast plains of South America; and other beds of different kind, but also of the same age, are known in India under the local name of *Kunkur*. The various accumulations of sand and marl partially filling up caverns, not only in England and Northern Europe, but in Sicily and other parts of the South ; and not only in Europe, but in Asia, North and South America, and even Australia, are of the same age. All appear to to be the result of the agency of the sea sweeping over a large tract of broken and loose rock for a long time, re-arranging the material occasionally by its own motive and carrying power, but very often receiving and carrying along by currents very large quantities of broken rock cemented together for a time by the ice in glaciers, and broken off from Arctic or Antarctic land to form icebergs. The time thus represented must certainly have been very great to account for the work done and the way it is done. The work must also have been completed very long ago, for wherever we now find the smallest indication of it, there must have been a complete upheaval of the whole land with this its last load upon it, and this upheaval must have been exceedingly slow, inasmuch as it has left nowhere any marks of that destruction that must have followed a rapid movement.

The gravels of the drift period, therefore, are

among the latest of the various groups of deposits completed upon the earth in those parts now above the water. Similar deposits may have been going on ever since, and doubtless are still accumulating in many parts of the ocean, but being formed in and by water, and therefore concealed from all observation during formation, the very fact of their being visible proves that they were completed long enough ago to allow of their subsequent elevation to their present level, often many hundred and sometimes not a few thousand feet above ·the sea. As also there is no evidence to justify the assumption, even as the vaguest probability, that the level of water in the ocean has lowered, we are bound to suppose that the slow upheaval of a few inches, or at the most a very few feet in a century, is the extreme average that can be assumed. In many parts of the earth newer beds than those of the drift period overlay them—thus the bluffs of the Mississippi, sea beaches in many parts of Northern Europe above the level of highest tides, the fine mud of the Rhine valley, all the principal river deltas in the world, and a vast accumulation of volcanic ash, lava, and other erupted matter— these, and the miles of coral reef still expanding widely in the South Pacific and Indian Oceans, are deposits even more recent than the drift of which we are about to speak.

To illustrate still further this important point, let us place in a tabular form the various deposits, so far as they are known, distinguishing those found in our own country from contemporaneous deposits abroad. This is indeed by no means an easy task, nor is it one than can be executed with absolute correctness, but enough is known to justify the following table. It must be understood that the larger groups, one to five, refer to the divisions of the whole period as nearly as possible, and the different sub-divisions in each are smaller groups that may have been contemporaneous, and that are not found generally on the same spot. The cavern deposits of England may, and probably did, occupy a long period; but some of those containing human remains certainly belong to the lowest or oldest part of the last principal group of the table, while others, whether with or without indications of the presence of man, have undergone successive additions up to the present time.

Tabular view of the relative position of the various beds and accumulations in England and abroad, during the deposit of which man is supposed to have existed.

BRITISH DEPOSITS.	FOREIGN EQUIVALENTS.
I. { *Peat beds,* with human remains. *Alluvial deposits* of the Thames, Mersey, and other river valleys, with buried canoes and stone hatchets.	I. { *Marine strata,* Temple of Serapis, near Naples. *Freshwater strata* with remains of ancient buildings, in Cashmere.

N.B.—The principal river deltas of the Old and New World, *e. g.,* those of the Nile, Ganges, Mississippi, &c., may be conveniently placed in this part of the table, although extending over a large part of the newer tertiary period.

BRITISH DEPOSITS.	FOREIGN EQUIVALENTS.
II. { Various marls, sands, and gravels, with recent shells, in different parts of England. *Shell marl* of Scotland and Isle of Man, with bones of Irish elk and other extinct quadrupeds. *Raised beaches* on the west and south coast of England, with occasional shells.	II. { *Volcanic tuff* of Naples, with shells. *Newer boulder formation.* Raised beaches, Sweden. *Loess* of Rhine Valley. *Tchornozem*, or black earth of Aralo-Caspian. *Regur.* Cotton soil of India. *Bluffs* and various deposits of Mississippi Valley. Newest marine deposits of Patagonia. *Newest gold gravels* of California and Australia. *Drift*, with bones of Dinornis, in New Zealand.
III. { GLACIAL BEDS, drifts, and boulder formation of British Islands, with remains of extinct and recent quadrupeds and recent shells. *Ochreous gravel* of Thames valley, and *Till* of Clyde valley and North Wales.	III. { GLACIAL DRIFT, gravels, and boulder formation of Northern Europe, Asia, and America, with numerous mammalian remains and *human remains* at *Abbeville*.
IV. *Pre-glacial deposits* of valley of Thames (Grays, Thurrock, &c.), with elephants' bones and shells.	IV. *Gold gravels* of Siberia, and some of the older gold gravels of Australia and California.
V. { NORWICH CRAG, with numerous bones of quadrupeds, many of them extinct species, and many shells, a large proportion of them recent (*human remains noticed by Mr. Frere'*) CAVERN DEPOSITS of the British Islands (Kirkdale, Brixham, &c.), containing numerous bones of extinct and of some recent quadrupeds, and *remains of human art.*	V. { *Sicilian limestone*, and caverns with mammalian (including human) remains. *French cavern deposits* (South of France). *Bone breccia* of Gibraltar. *Australian deposits* with extinct marsupials. *Brazilian cavern deposits*, with extinct quadrupeds. *Kunkur* of India. *Newer Pampæan formations* of South America, with gigantic edentates.

It requires a good deal of consideration before one can even imagine the myriads of revolutions of our globe that must have been required merely to perform the surface operations which we are able to examine, and which we know to be actually more modern than the drift. And yet, during the drift period itself, a large proportion of its animal and vegetable inhabitants were the same as now, and the change, whatever it may have been that has affected the races then living, does not seem to have been sudden. Some species have indeed died out, and have been replaced by others better able to endure the altered state of things; but the list of these lost species can hardly be said to have extended beyond the quadrupeds and birds. Occasionally we find in our seas, or on the shore, species of shells extremely rare in a living state, and common enough buried with the drift deposits; but these are generally either Arctic forms, brought to our neighbourhood when the climate was colder, or Mediterranean and South European species (from the shores of Spain and Portugal), which remained behind in small number through the cold period.

The common gravel of the south and south-east of England is made up chiefly of rolled flints from the chalk; that found in the north is either granite or other local rock broken away from the mountain country to the west and north-west. In the valley of the Clyde is a deposit of clay mixed up with rounded boulders, often of large size. In various caverns are sand and loamy or marly beds, carried in by water, and forming an even floor, mixed up with bones and teeth, and often cemented over by a coating of carbonate of lime. Wherever the drift occurs, it is, in the proper sense of the word, local—that is, in patches, limited to a comparatively small area; any two patches not having much to do with each other, however near they may be, unless there is evidence of their having been separated from one another since their deposit. These drift beds are sometimes of freshwater origin, consisting of sand and loam, and contain snail shells and the shells of animals inhabiting pools and rivulets; but more frequently they are composed of rolled fragments, evidently water-worn by tidal action, and heaped together with sand and clays, and occasional huge blocks, without regard to order or specific gravity. There are sometimes found mixed with the sand remains of marine animals, either shells or bones of fishes; and there are not wanting many instances in which the accumulation of gravel is nothing more than an ancient sea-beach.

The gravel beds lie irregularly over most of the known and named rocks of geologists. So irregular are they, that in geological maps they are actually left out altogether, as either too imperfectly made out to justify insertion, or so completely obscuring the other rocks that the order

and arrangement of these latter would be lost
sight of if their upper coatings of gravel, often
forming the true subsoil, were represented. It is
only in special geological maps, constructed with a
view to these superficial deposits, that they can be
inserted with advantage; and yet for certain practical purposes, especially those relating to agricul-
ture, and also for many in which engineering and
architecture are concerned, a knowledge of them is
absolutely essential.

These beds are so different from the underlying
rocks, such as the chalk, lias, coal-measures, and
others well known in various parts of the country,
that they may be regarded as exceptional in their
character, and they are considered to have been formed
in a sea loaded with icebergs drifted down from the
circumpolar seas. That most of the material has
been conveyed for some distance by the sea, there
is no doubt; and in many cases the surface of the
hard rocks on which gravel beds lie, or which are
very near them, is smooth and scratched, as if
by dragging over them some heavy load of rough
material. Glaciers rapidly formed along a wide
extent of circumpolar land at a high elevation,
and frequently broken off to form icebergs, which
again, with their heavy load of stones and gravel,
accumulated on the sides and slopes, as they do
now in the Alps and other lofty mountain ranges,
offer the best, if not the only, explanation of the
conditions under which the boulder clay and coarser

gravel could be formed. The gradual wearing away of cliffs, and the removal and re-accumulation of the broken material on a line of coast, is another process that must have been going on contemporaneously, and may have formed the more regular and finer banks of gravel and sand, while the silt, stones, and mud brought down by torrents and left behind on the low ground, through which the waters of a torrent have found their way to the sea, would produce the freshwater sands and marls belonging to the same course of events. In the times, then, of which we are speaking, there must have been a large district, either a group of islands or a continent, extending much further to the south than the present Arctic land; and on this land, part of which may have extended to the latitude of Central Europe, the climate must have been excessively cold, almost Arctic, owing to inhospitable mountains of ice everywhere advancing towards the sea. Still further to the south were numerous islands, if not continuous lands, covered with vegetation, and peopled with various tribes of quadrupeds and birds, and probably (as we have seen) not without human inhabitants. Such islands occupied chiefly those spots now at a great elevation above the sea, the lower plains being submerged, though probably not so deeply but that the icebergs were stranded upon them. The central plain of Europe, a large part of Asia north of the great Himalayan chain, an extensive tract of North America, a broad strip of South America east of the Andes, from the river La Plata southwards, and a part—no one can at present tell how large a part—of Australia, were all then under water. It was on the portions of the ocean floor of that day, defined as above, that the great deposits of the drift period took place. The floor is since raised, and the ocean has left it; some of the ancient land is now submerged, while other parts form mountain-tops; the sea now brings warm water instead of ice to our shores, and the drifted icebergs deposit their load in the middle of the broad Atlantic, instead of on shoals a thousand miles to the east or west. The climate has changed entirely; the vegetation and the inhabitants of land and sea have also changed more or less; but man, with his power of modifying his habits to climate, has remained; and many animals of lower organization, moving freely in water, and able to find for themselves those conditions of climate and food that are favourable, have also remained as his contemporaries.

It will be interesting to inquire what were those animals associated with man during this period, which is so historically remote, but geologically so very recent—animals many of which have since died out, leaving no direct and unaltered descendants on the earth. They include some of many kinds, large and small, belonging to various countries; but it will not be possible to allude here to

more than a few of the more remarkable and characteristic varieties.

In the caves, both of England and Germany,
bones of bats are found buried with other bones
under the stalagmite, as well as amongst the superficial mud. Two species at least are made out—
one of them the " great bat" of English naturalists,
still living in similar places; and the other the
horse-shoe bat, also living still. In these curious
animals, limited in distribution to sheltered and
dark places, there has been, then, but little change.

Bears, of at least three kinds, certainly inhabited
the land during the drift period; the brown bear
of North Europe, which is said to have lived in
Scotland less than a thousand years ago, had
already been introduced, and was accompanied by
two other species, both since extinct—one of them
smaller and less fierce, the other much larger, and
more resembling the grisly bear of North-Western
America. The great cavern bear, as the larger
extinct species is called, must have equalled in size
a large horse; and though certainly very powerful,
and from the structure of its teeth and extremities
able to defend itself against enemies, it probably
fed more on vegetable than animal food—in this
also resembling the grisly bear.

The badger, the polecat, and the stoat flourished
with the three species of bear above described, but
seem not to have since undergone any change,
resisting the alteration of climate and the gradual

increase of the human race better than the larger animals of approximate habits. Bones of all of them are found buried in caverns with those of extinct races. The otter was another existing species which at that time had been introduced into Europe, and has not since been destroyed.

Wolves abounded during the deposit of the drift, and cannot yet be said to have passed away, except where the cultivation of the land has rendered their existence impossible. The same species now common throughout Northern Europe, then ranged over England also, and numerous individuals have left bones and teeth in the caves by the side of those of the bears. The same may be said of the common dog, of which there seem to have been many varieties, and of the fox.

The hyæna is an animal combining many of the peculiarities of the canine and feline races. It includes species less destructive than many, the animals seeking rather dead carrion than the living prey. The teeth of the hyænas are admirably adapted to gnaw and crush bones, and the muscles of the jaw point out this as an important habit. At present animals of this kind are confined to Africa and the parts of Asia adjacent. Two well-marked species and one other are known; one only, the striped hyæna, inhabiting Northern Africa and Asia, and the others found in South Africa, near the Cape. One very remarkable extinct species, more like the spotted hyæna of the

South than the striped species of the North of
Africa, has left abundant remains in caverns,
among the fossils of the drift period. This last
species has been found in the principal caverns of
England, Germany, France, and Belgium, and also
in the unstratified drift of the same period. It is
singular enough that the specimens from the latter
localities are in a tolerably perfect state, not broken
or gnawed, while those from caverns almost always
show marks of having been gnawed and broken by
the teeth of their cannibal associates.

The hyæna of the caves was an animal equally
remarkable for size and strength, exceeding very
greatly that of the spotted hyæna of the Cape,
and attaining dimensions even larger than those
of the largest tiger. The number of individuals
of this species whose remains have been found in
the English caverns, is not only extremely large,
but they belong to animals of all ages. It seems
clear that the caverns were the dens and hiding-
places of these savage brutes, who dragged thither
the deer or other prey they found while prowling
about at night, and often, in default of other
victims, fed on their own companions or offspring.

It is a common error to suppose that the
larger carnivora are only met with where the climate is tropical, or warm and damp, and that a
large development of animal life—especially of the
larger races—is always accompanied by a vigorous

vegetation. The contrary might almost be asserted as the usual condition in nature. Thus in South Africa, where are the largest known herds of wild animals, both carnivorous and herbivorous, they exist in tracts of country scantily covered with vegetation, and in Brazil, where the vegetation is so enormously multiplied, there seems hardly room for large animals. So also on the northern side of the great mountain chain of Asia, and in the colder parts of North America, the larger carnivora are represented by lions, tigers, lynxes, and jaguars, with many other feline species.

In ancient times, during the arctic climate of the drift period, a great cavern tiger ranged over England, and all that then existed of Europe, accompanied by another species of tiger of smaller size, a leopard, a wild cat, and some other of the larger feline animals, besides a very remarkable genus (*machairodus*) now altogether lost, as large as the cavern tiger, and provided with weapons rendering it if possible more formidable. Of these, the cavern tiger seems to have resembled the jaguar in its proportions, but was stronger and larger in the limbs and paws; the others were fully as large as the existing tiger, but all were slightly different, except, indeed, the wild cat; whilst the extinct *Machairodus*, with its peculiar canine teeth, shaped like a sword (whence the name is derived), and having a serrated edge, was not only larger than any of these, but seems to have been, in spite of

some bear-like affinities, a yet more destructive creature.   It ranged over the whole of the northern
temperate land, from England to India; its remains
having been found, though rarely and at intervals,
across the two continents of Europe and Asia.
Several species have already been made out.

Accompanying this striking list of ferocious and
carnivorous animals, we find a corresponding group
of vegetable feeders serving as their prey.   The
beaver, and a species of rodent of very large size
closely allied to it, have been found, together with
a large group of the more minute gnawing animals;
but with these were also the gigantic races of
which the representatives now are the Indian and
African elephants, the rhinoceroses and hippopotamuses, and the tapirs of the Old and New World.
From a period long antecedent to that of the drift,
and down to a comparatively recent date, elephants
and rhinoceroses seem to have abounded in northern
latitudes, both in Europe and America, ranging
even to the borders of the Arctic circle, wherever
a tree vegetation could live.   So lately have they
there died out, or so long have their fleshy and
soft parts been embalmed in ice in those countries,
that in the beginning of this century a perfect
elephant was melted out of the icy cliffs of the
River Lena, in Siberia, which must have been suddenly entombed when in full health, and while
living in a natural state in such a climate as still
exists in the same parallel of latitude in Western

Europe. The carcases of a rhinoceros and other animals have been seen also, but these have been left where they were seen, while the skeleton of the elephant, the skin and some of the hair, and the contents of the stomach, were sent to St. Petersburg.

The bones of elephants, rhinoceroses, and hippopotamuses have been found very generally, both in cavern deposits and in the gravel of the drift period, so that no doubt can exist as to more than one kind of each having ranged over the whole of the land during the deposit of the various beds of gravel. They must also have been locally abundant to account for the very numerous bones distributed wherever gravel appears or where caverns exist.

Of these one of the elephants (the mammoth), grew occasionally to enormous size, much exceeding that of the largest known Indian elephants, and was provided with tusks proportionably elongated and greatly curved. It may serve to give some idea of the abundance of these animals in England, if we mention the opinion of Mr. Woodward, that from a bank off the little village of Happisburgh in Norfolk, upwards of two thousand grinders of the mammoth have been dredged up by the fishermen within thirteen years. And even this is by no means the richest locality, as all along the east coast of England, from Essex to Norfolk, the teeth and tusks of elephants seem to be among the com-

monest fossils dug up, while in the cliffs bor-
dering the frozen Arctic seas, the supply of
tusks was at one time, and for a long while, so
considerable as to render the importation of ivory
from recently killed elephants altogether unnecessary. It is remarked, indeed, by a traveller
who published an account of the Lae-chow Islands,
on the north-eastern coast of Siberia, that one of
these islands is little more than a mass of elephant's
bones.

An equally gigantic and somewhat more massive animal than the elephant, but very closely
allied to it—the mastodon—is now extinct, but
during the drift period trod the wastes and fed on
the tree vegetation of Northern Europe, Asia, and
America. The animals of this genus seem to
have been more widely spread than the elephant,
ranging from the tropics, both southward and
northward, almost to the Polar Seas, and reaching
back much farther in geological time. They were
provided with a long proboscis, and were possibly
more aquatic in their habits than elephants, ap-
proximating in form and habits to the hippopotamus, species of which have also been found
embedded in the same deposits. Of these one was
larger and another much smaller than the species
now inhabiting the Nile.

A well marked species of rhinoceros of the two-
horned group, provided with horns of remarkable
size and strength, has been found both in gravel

and caverns, and a complete skeleton was met with in a natural fissure at Wirksworth, in Derbyshire, laid open by mining operations. Other bones of the same, and some probably belonging to a distinct species, have been found in many parts of Europe and Northern Asia.

Two species of the genus *Equus*, one equalling a middling sized horse and the other a zebra, have left remains in caverns and gravel, and are accompanied by bones of animals assisting to fill up the wide interval at present existing between the hippopotamus and the hog. Several of these, of various sizes, and the wild hog itself, evidently ranged throughout the northern countries during the drift period.

Of ruminating animals, there are numerous remains in the caverns and in the gravel of the drift period. Two species of camels, a gigantic musk ox, and a giraffe, were the prototypes of many existing and well known groups. Deer of all kinds were then, as they still are, represented by many and greatly varied types, and of these the great-horned Irish elk, as it is called, seems to have been the most remarkable for its dimensions, and from the enormous expanse and width of the horns, which in some specimens exceed fourteen feet from tip to tip, and weigh upwards of eighty pounds. The animal in certain respects resembled the fallow-deer, and in others the reindeer, and was not really an elk. Its horns must have fallen off and been reproduced every year, and exhibit proof of their enormously rapid growth in the deep grooves left for the passage of the blood vessels which conveyed the material of which they were formed.

A true elk, a rein-deer, and a gigantic fallow-deer and red-deer, have all left bones and teeth in caverns and gravel, and these are occasionally accompanied by remains of roebucks, antelopes, and goats. The *Aurochs*, a large bison still roaming in the wild forests of Lithuania, and a gigantic ox, the *Urus*, scarcely inferior to the elephant in size, both lived in England as well as Northern Europe, about the commencement of the Christian era, and had lived in the same district during the whole of the drift period, together with a smaller species, about the dimensions of the ordinary domestic cattle.

Whilst these were the inhabitants of Northern Europe, South America was provided with several gigantic representatives of the existing tribes peculiar to that country. Thus, instead of, or in addition to, Armadillos, we find bones of the *Glyptodon*, covered with a coat of mail six feet in length, its vast bulk supported on massive columns, with bases of corresponding magnitude. With this we have the *Toxodon*, also a large animal, combining some of the peculiarities of the elephant and the beaver; a gigantic llama, with a body as large and massive as that of the rhinoceros, and if last, certainly not least, the megatheroid group— the most remarkable of all, including sloths larger

than the largest elephant, whose habits were not
unlike those of the sloths of the present day,
except that instead of climbing the trees, their
enormous strength enabled them to pull down the
giants of the forest, and strip them of their leaves
at pleasure.

The caverns of Australia, like the Pampas deposits of South America, contain the remains of an
ancient fauna, strangely typified by the existing
inhabitants of the country. As in the latter
country the sloths and armadillos appear to be the
dwarfed successors of more gigantic animals, similar
in structure and habits, so in Australia the peculiar
characteristic of all the indigenous quadrupeds—
their marsupial structure—(the mother bringing
forth her young in an immature state, and preserving them in a bag or pouch till they are ready
to provide for themselves) is retained in a group
of gigantic prototypes of almost all the chief
natural groups. In New Zealand, where no quadrupeds of any importance existed at the time of its
discovery, there had been in like manner a gigantic
race of wingless birds, of which the modern apteryx
and the recently lost dodo may serve to give an idea.

Everywhere then we find occupying the land of
the drift period, when the gravel was deposited
and the caverns filled, important groups of quadrupeds, many of them of gigantic size, but all resembling pretty closely the present inhabitants of
parts of the earth not very far removed. Thus we

do not find in Northern Europe and Asia the remains of sloths and marsupials such as now inhabit South America and Australia, but of elephants, rhinoceroses, and hippopotamuses, of lions, tigers, and leopards, and of great deer and oxen—all of which are still more or less accurately represented there in a living state. Climate appears by no means to have limited the distribution of these creatures in former times, as in parts of the world certainly colder then than now are the remains of numerous species of which the nearest allied species are only met with in much warmer countries at present. That a similarity of typical character pervades the inhabitants of the same district, and seems to have done so through a vast period of time, is a point well worthy of remark, and cannot but have important reference to the laws, whatever they may be, which govern the succession of species.

While, however, the larger land animals were thus different, the inhabitants of the water, and generally the smaller and lower organized races of all kinds of the drift period, seem to have altered but little, if at all, in passing through the changes that have taken place from that time to this, since only a few shells, then common, are now even rare, and none probably have so far changed as to be worthy of being called new species. The same is the case as far as the evidence goes with the vegetable world. Notwithstanding, therefore, the

lapse of the long series of ages required to introduce so many important modifications in the
animal world, it is as nothing compared to those
other periods contemplated by geologists, during
which almost every species, both vegetable and
animal, has been changed over and over again.
The drift period is the first of those numerous successive steps by which we are conducted in tracing
the history of the past, and being the first, and
that which involves fewest and smallest changes, it
is equally important in guiding our judgment, and
interesting, as showing better than any other, the
method and prevailing law of nature in the progressive history of creation.

We have seen in a former chapter that the drift
period—so modern in reference to geology, so
ancient in reference to human records—was marked
not only by races of quadrupeds different from the
present inhabitants of land, but also by tribes of
men who, in their habits of life, as judged of by
their weapons, differed little, if at all, from those
tribes inhabiting Western Europe two thousand
years ago, and equally little from those found in
America two centuries ago, and in many islands in
the Pacific and in the northern and unsettled parts
of Australia at the present day. Now, as then, the
human race, in its power of resistance to cold and
heat, hunger and thirst, and extremes of all kinds,
possessed an elasticity due no doubt to intellectual
pre-eminence, which distinguished the most savage

tribes and rendered them superior to the most powerful and intelligent of the unreasoning animals. They remained and held their ground, whilst all, or most, of the others were forced to give way to changing climate and altered conditions of food. They continued to rule long, perhaps without much civilization, but always the lords of creation, until the early and uncivilized tribes were driven out by others having more cultivation, and therefore greater power. They have left behind a few stone weapons found in the caverns and gravel, buried with the bones of quadrupeds of their day, now extinct. Hitherto the subject is new, and the discoveries that can be depended on are few and inconsiderable; but we may look forward to the time when other indications than stone hatchets, knives, and arrow heads will be found, and when the bones and skulls shall testify to the peculiarities of the race. Perhaps indeed many such have been picked up already, and lost again, owing to the total absence of preparation, even of scientific observers, to admit them; but now, that the possibility is recognised, we shall soon see the evidence grow around us.

It is curious and interesting to find in this way another important department of human knowledge —archæology—taking its earliest facts and conclusions from geology, adding thus to the large number of sciences based on the history of the earth and its contents. The facts brought forward by geological investigations concerning gravel and caverns are at present the only sources of information for determining the antiquity of our race.

There appears to be nothing in the Nilotic, or cuneiform inscriptions of the most ancient date, nothing in the oldest hieroglyphics of Egypt, nothing in the sacred writings considered as a history of the human race, that points to the existence of an ancient people widely spread over the earth in both hemispheres, living in a savage, or at best half civilized state, but capable of manufacturing arrow and spear heads and knives, roughly hewn out of hard stone, and in Egypt, at least, brick and pottery. Far anterior in time to all ordinary records, to cyclopean constructions in stone, or to those pictured and sculptured stones, however uncouth ; their implements and weapons found in many countries, in caverns, in gravel beds, or in river mud ;—these tribes are, for the most part, dissevered from the oldest that show relation with actual history, and carry us back to the period when the last races of large quadrupeds ranged over the earth. Sculptured by beings who must have shared the earth with the cavern bear, hyæna and tiger, the elephant, mastodon, hippopotamus, and rhinoceros, the great Irish elk, the megatherium and glyptodon, the gigantio kangaroos and wombats, and the dinornis, and numerous other strange quadrupeds and birds, these implements, simple and rudely formed as they are, still show a degree of ingenuity, and indicate a certain amount of intellectual cultivation, compared with which the cleverness of the monkey, or any other unreasoning animal, however perfect in instinct, is not to be brought in comparison. There is enough in what is already known to stimulate curiosity, to hint very obscurely at the earliest condition of the uneducated human being, and at the same time to point to a lapse of time marked by geological epochs, instead of years or centuries. How long this state of things lasted, and by what steps men grew out of this infancy into the partial and obstructed cultivation of the Chinese on the one hand, or the hitherto unchecked progress of the Saxon races on the other, or what intermediate gradations are represented by the Celtic tribes, the Aztecs, the Egyptians, the North American Indians, the South Sea Islanders, the Boschmans, or the indigenous Australians, there are no means yet of determining. We cannot even make out whether this early race was destroyed, or nearly so, to make way for a newer one of greater intellectual activity, or whether the newer arose out of the older obeying the great law of intellectual progress.*

---

* The latter part of this chapter was originally published in the National Review, vol. x. p. 279.

XIII.

ORIGIN OF ROCKS AND METAMORPHISM.

Rocks supposed to have been fused—Absence of evidence on this head—Chemical geology—Cleavage—Kind of rocks that cleave —Nature of mud deposits—Mechanical production of cleavage in soft materials by pressure—Investigation by aid of the microscope—Granite—Vesicles and cavities in rocks occupied by water—Various conditions of occurrence—Composition of granite and of the crystals of which it is made up—Differences of composition of granites from different districts—Oolitic structure—Its mechanical origin.

FROM the time of Werner to the present day, there have been discussions among geologists as to the origin of rocks, and extreme difference of opinion has been entertained as to whether certain rocks, which are evidently not the direct result of aqueous action, are to be regarded as truly igneous, that is, due to the agency of intense heat, or only *metamorphic*, by which is meant, changed by heat or chemical action from original aqueous deposit. Granites, and the large class of rocks of which granite is the representative (technically called *porphyries*), crystalline limestone, and quartz rock have been generally described as cooled down from igneous fusion, while slates, gneissic rocks, and some others, retaining marks of mechanical origin in their planes of bedding, but having either no fossils, or fossils only in a most obscure form, have been thought to have undergone great change and partial fusion in consequence of their close vicinity to rocks in a perfectly fused state. The known condition of recent lava, and the alterations caused by it on many rocks, have long been considered amply sufficient to justify the assumption that all crystalline, or greatly altered rocks, are so because of the action of heat upon them. For some time past those theorists in geology who believed that the earth was still in a state of igneous fusion at a comparatively small depth from the surface, and that the alteration of rocks was caused by their being brought within the influence of this heat under great pressure, have been by far the most popular and numerous, insomuch that the opposite view has almost been lost sight of.

In some respects, no doubt, the old battle of the Neptunists and the Vulcanists has been fought out and forgotten. No one now, with Werner and his immediate followers, would argue, in the face of known and recorded facts, that *basalts* (as ancient lavas are called) are chemical precipitates from water—for the identity of basalt with lava recently poured out from volcanic vents has been experimentally proved. No one, on the other hand, attempts to carry the igneous theory so far as to deny the vast importance of chemical agency independently of heat, or assisted only by such moderate and equable temperature as is known to belong to all considerable depths beneath the earth's surface.

With the advance of geology and chemistry, each in its own direction, there has arisen a department of science spoken of sometimes as " chemical geology." It is not, in the nature of things, likely to prove so popular as some other branches of the subject, especially the study of the extinct races of animals and vegetables.   Still there are matters of general interest treated of by chemical geology which seem to admit of popular elucidation, and a few of these we propose now to discuss.

Cleavage.—In all rocks that are not absolutely crystalline, there is a kind of parallel structure, more or less easily recognisable. It might result from the way in which the materials of the rock were accumulated, after being for a time held in suspension in water, just as at the present time layers of mud are deposited parallel to each other at the mouth of the Rhine, the Nile, or any other great river. It might, however, be the result of some change that has taken place since the original accumulation was made, involving subsequent modification of a mass originally heaped together without order.

In nature it is chiefly clayey beds that show this second kind of parallel structure, and in them, in its most perfect form, it produces roofing slate. The varieties, both of slate and of imperfect slates

called schists, as well as the rock called gneiss, are
similarly formed. The term *cleavage* is used to
denote the structure in question. Many substances
(amongst the rest wax) may be made to assume
cleavage by simple exposure to pressure.

All beds produced by the accumulation of mud,
whether at the mouths of rivers or in open seas,
are deposited in successive parallel layers of the first
kind, and often contain fragments of animal remains. They are chiefly clayey or sandy substances,
brought down by rivers, or washed from cliffs mixed
with a certain proportion of carbonate of lime. A
carefully observed boring, made to a depth of 1800
feet, in Amsterdam, through the accumulations of
Rhine mud which are there the only deposits
known, showed in the first 232 feet eleven beds of
clay, whose total thickness is 153 feet, and eight
beds of sand, whose thickness is 79 feet; but
below this there was nothing but sand. At a
depth of 138 feet from the surface, the clay
abounded with extremely minute and simple forms
of vegetable life belonging to the *Diatomaceæ*, from
one-third to half the mass being organic. In other
river deltas, as in that of the Nile, borings have
also shown the existence of alternate deposits of
sand and clay, and here also the clay contains
diatoms and the calcareous framework of foramini-
feræ. Clay slates consist of a combination of silica
with alumina, and various earths and metallic
oxides, very similar to that which has been observed

in clay bands recently deposited, but the clay must have been of a peculiar kind, and has not been proved to contain fossils.

The production of cleavage or the superinduced parallelism of the particles of a rock, is not unfrequently seen to be altogether independent of the stratification or original deposit, though sometimes the two are themselves strictly parallel. It is rare in any but rocks of great antiquity to find good specimens of cleavage at all, and even then they are only discernible when there is independent evidence of a large amount of pressure, both vertical and lateral, having been brought to bear upon the deposit.

In cleavage, unlike ordinary stratification, there is no practical limit to the splitting of the rock in some one direction (thence called the plane of cleavage). As this is the case also with crystalline minerals, many of which when pure cleave very readily in some directions, but not at all in others, the first idea of the cause of cleavage in rocks very naturally was that crystalline action had gone on in the mass, and had thus produced one of its very important effects. As, however, no other essential character of crystals could generally be traced, this theory was not quite satisfactory.

Mere pressure, when sufficiently great, was found by experiment to produce a phenomenon so exactly resembling cleavage that this has been put forward as a sufficient cause. No doubt most of the clay-

s

slates and other rocks showing this peculiarity of structure have undergone vast pressure when buried beneath a load of rock and water. This has been increased when, besides the pressure from above, a force acting from below has compressed or broken the overlying mass, or when, as has often happened, the elevatory force has even been sufficient to overcome the inertia of the overlying heap of material, and actually to lift it bodily upwards. Pressure is obtained, perhaps, on an equally grand scale when, owing to altered conditions of the rock in the interior, from chemical causes, that contraction has taken place which must have accompanied the drying and consolidation of all rocks, and of which there is so much evidence in the cracks and fissures penetrating those that are not squeezed together into a much more compact mass than their original mode of deposit would admit. Lastly, the expansion of rocks, when carried down to greater depths and into higher temperatures than those in which they were formed, must have tended to produce the same result. From many causes, therefore, it is evident that enormous pressure has acted, especially on those rocks which are of oldest date and which bear marks of having been carried into the deeper recesses of the earth to elaborate their structure.

The microscope in the hands of Mr. Sorby has revealed something of the origin, while exhibiting with accuracy the structure, of cleavage rocks.

It appears, on careful examination, that in two rocks, of which the materials are the same, but one of which shows cleavage and the other does not, the particles are really differently arranged, the appearance corresponding to what would happen if the rocks had been squeezed flatter in a direction at right angles to the cleavage, and pulled out in the direction of that plane. Some curious proofs of a change of this kind are seen in the oval form of certain green lumps common in clay rocks, and still more in fossil shells belonging to the slate period, all of which, in the slates or other rocks exhibiting true cleavage, are distorted, although, in the parts of the same rock where there is no cleavage there is no distortion. The clay, however, in slates is not exactly of the same kind as that now formed in water, but rather seems to be made up of exceedingly minute plates of mica, and in this ultimate sense the particles, though not the masses, are crystalline. The idea of the formation is that the mud from the decomposition of the felspar of granite was changed into mica under the influence of water at a high temperature, and deep below the earth's surface; that after this, owing to the action of the causes already indicated, the dimensions of the rock have been changed without compression of the whole mass or further crystalline action, and that the water being squeezed out, and the mass to a certain extent reduced from plasticity to a solid, there were produced at the same time the two extreme kinds of cleavage—that illustrated in the fine roofing slates of Wales, Scotland, Cornwall, and elsewhere, and that occasionally shown when rocks split indefinitely parallel to natural joints or crevices produced in the rock, independently of true cleavage. A fair illustration of the nature of cleavage in its characteristic form is produced mechanically by mixing carefully numerous scales of micaceous oxide of iron with pipe-clay, and afterwards, while the whole is plastic, subjecting it to pressure and great strain in one direction. If after this the specimen is baked so as to render permanent the new arrangement of the contents produced by the pressure, a little examination will show that the scales of iron oxide have all taken a distinct direction, parallel to the elongation of the clay. A mass of any material having an ultimate crystalline structure is, however, sufficient to produce the effect, without the presence of foreign bodies.

Granite.—To most geologists it would be a startling assumption if granite, and the large group of rocks of which granite is the best known example, were declared to be in any sense aqueous rocks. Undoubtedly crystalline, and made up of crystals of quartz, with felspar, mica, hornblende, or other double silicates, all in a crystallized state, and accumulated together; showing little indication of mechanical origin; and not containing fossils or remains of animal or vegetable organization, it certainly does not at first seem possible to speak of such a substance and water as having anything to do with each other.   Still, the microscope has revealed facts concerning rocks of this kind which go far to prove that, if not Neptunian in the old Wernerian sense, they are at least not without some very curious and satisfactory proofs of water having had so much to do in their formation that a part of this element has actually been locked up in their substance, and allows us to calculate at what depth and temperature the whole mass was formed.

By obtaining sections from one hundredth to a thousandth of an inch thick of various crystalline bodies, it has been found that numerous cavities exist in almost all, but that these cavities vary greatly according to the circumstances under which crystallization took place.   They are usually occupied by fluid or gas, and by ingenious devices the presence of water has been detected in many, the fluid not being chemically combined, but mechanically enclosed within the cell-walls of the cavity.   Quartz seems particularly liable to contain water under these circumstances; and the cavities are sometimes quite large enough to be seen by the naked eye, in which cases the fluid can be extracted and examined.   Besides these large cavities, there are often innumerable smaller ones discoverable only under the microscope, although the result of their existence is visible in a white appearance of the mass.

Now, it is a curious and important fact that
minerals obtained in a crystalline state by direct
cooling from igneous fusion, also contain cavities;
but these are generally either full of some transparent glass derived from the cooling mass, or are
quite empty, or else are occupied by gas or vapour.
Sometimes, also, they contain small crystals. When
the larger cavities in such rocks become filled subsequently with crystals, these crystals contain fluid
cavities—thus marking the presence of water, and
proving that it circulates through partially cooled
igneous rocks, and especially in the cavities with
which they abound. The rock mass, therefore,
cooled down from igneous fusion, contains empty
cavities, or spaces, without water; but whatever
has formed afterwards in these spaces contains
water.

Granite rock consists of crystals of quartz,
felspar, mica, hornblende, &c., or crystals of some
of these or of allied minerals, embedded in a base
of one of them. The quartz of granite contains
so many cavities filled with water that the actual
bulk of the water thus contained is sometimes one
per cent. of the bulk of the rock. The fluid
cavities are distributed through all parts of the
mass. The quartz of granite, however, also contains stone and vapour cavities; but the coarser
the granite, the smaller and less distinct are these.

It results from the observations made in this
way that granite must have been formed like some

modern volcanic rocks, no doubt at a high temperature, but still at a temperature not much greater than is required for the fusion of lead, and under the pressure of many thousand feet of overlying rock. Now, as it requires an enormously greater heat at the ordinary pressure of the atmosphere at the surface to fuse such rock, it is quite clear that granite cannot be regarded, as it often has been, as a result of complete fusion of the whole mass while far beneath the surface; and Mr. Sorby, to whom we are indebted for this investigation, says—"The proof of the operation of water is quite as strong as that of heat; and, in fact, I must admit that, in the case of coarse grained highly quartzose granites, there is so very little evidence of igneous fusion, and such overwhelming proof of the action of water, that it is impossible to draw the line between them and those veins where, in all probability, mica, felspar, and quartz have been deposited from solution in water."*

There is a great difference observable in granites from different localities. Thus the granite of the Highlands of Scotland appears to have been formed under a much greater pressure than those of Cornwall, or else at a much lower temperature. Assuming that the temperature is more likely than not to increase as we descend far below the earth's surface, and that the law of increase is not

---

* Quart. Journ. of the Geological Society, vol. xiv. p. 488.

different from that observed in deep mines, it would seem that all kinds of products formed within the earth, and therefore under pressure, whether they are completely elaborated under pressure, as the granites were, or partly educed while the pressure was removed during some great volcanic outburst, must indicate the mode of their formation by their structure, and unless actually in a state of fusion, must be so affected by water as to contain it in the cavities of crystallization. Vapour of water, thus rising through all rocks, and everywhere present at great depths, must also reach and pass through all overlying sedimentary rocks, and help, no doubt, to produce those modifications in them that have often puzzled geologists.

It must not be supposed that the water in these cases is always pure. It contains often chlorides of potassium and sodium, sulphates of potash, soda and lime, and sometimes free acids. In these cases the phenomena of the cavities are often modified, but the general results as given above are not changed.

Water also in these cases cannot be said to act by dissolving igneous rocks when mixing with them. The rock rather seems to dissolve the water, either chemically, the mineral becoming a *hydrate*, or physically, in the manner in which liquids dissolve gas, actually absorbing a part of the vapour of water into its substance. Under these circumstances, an intimate mixture may take the place of partially melted or softened rock and water at the same temperature, and it is, indeed, highly probable that in this way many rocks have been formed, just as quartz and felspar have actually been produced artificially by a French chemist, M. Daubree, in a manner which substantially confirms these views.

Oolitic Structure.—Many of the limestones of the middle of the three great series of formations which compose the earth's crust are remarkable in England for having a peculiar structure, the particles of the stone being small and round, like the roe of fish. This structure is well seen in the Bath and Portland stones, both very extensively used for building purposes, and is so manifest and characteristic that many of them are locally called roe-stones. In geological language the corresponding Greek word, *oolite,* is preferred, and oolites are so much more common in rocks of the age referred to than elsewhere, that we read in geological books of the oolitic period, so that it might be imagined that this condition of the stone was confined to the rocks in question. Such is not at all the case. Limestones thus oolitic, or made up of small rounded grains visible to the naked eye, occur in rocks of all ages—they are Silurian, Devonian, and carboniferous, as well as secondary, and not unfrequently, also, they are observed in tertiary beds.

It is well known that some limestones exist in

the state of chalk—fine grains adhering together
and forming a very soft stone ; others as oolites in
the large grains just described ; others, again, as
hard, compact, fine-grained masses, not at all
crystalline ; while others form more or less perfect
marble. Why these differences exist, and what is
the nature of the change that some of them have
undergone, is a curious question in geology. With
regard to the oolites, a short explanation will be
useful.

Almost all the little grains or egglike particles
which make up a piece of common Bath or Portland oolite, are compound bodies, formed separately,
and each having its nucleus of some very minute
organic body, such as a fragment of shell, or
other hard substance. Round this nucleus are
numerous layers of the finest lime-mud, mixed
with a little fine clay. In most cases the arrangement seems strictly mechanical, and is supposed to
have been induced by a slight ripple motion in
shallow water loaded with carbonate of lime, which
is thus deposited in exceedingly minute particles
on the fragments of lime sand (broken shell and
coralline) which cover the shore. The slight mechanical disturbance of the water is believed to be
sufficient to keep the carbonate of lime from set-
tling in a solid and compact deposit on any larger
body, and confine it to the minutest grains. If
this is the case, and the oolites have been thus
formed, the part of the sea where they were pro-

duced must have been in a state of slow descent, and this indeed agrees well with the nature of the fossils found so abundantly in the oolitic rocks. There are no circumstances apparently so favourable for the accumulation of fragments of shells, and the host of animal remains found in rocks, as when a coast line is undergoing slow depression, for the range of marine life is then constantly shifting, and often enlarging, by the continual addition of new shallow tracts. Sir Henry de la Beche noticed something of this kind in Jamaica, where the tidal range is small, but the waters are much charged with carbonate of lime.

By degrees, and under circumstances peculiarly favourable, the grains coated in this way have sometimes continued to increase till they are as large as peas, thus forming the rock called *pisolite* or pea-stone, in contradistinction to *oolite* or roe-stone. The pisolites are not often in thick beds, but in other respects they agree well enough with oolites.

When the carbonate of lime in the water is small in quantity, and does not accumulate in the way just described, it acts as a cement, and often fastens together sands and other material. Not unfrequently a deposit of iron goes on under similar circumstances, and thus have been produced those extensive and rich beds of oxide of iron found in the oolites, and of late abundantly worked in England and France.

XIV.

NEW DISCOVERIES IN IRON ORES.

*Wide diffusion of iron ores—Different kind of ores—Oxides and
Carbonates—Resources of England in 1851—Increased resources since that date—Composition of the common ores found
in the lias and oolites—Account of the Cleveland iron-pro-
ducing district—Its produce, and probable extent of its yield
—Other sources of supply from secondary rocks—French iron
ores—Silesian ores—Value of association of iron and coal to
England.*

NEXT to coal, nothing in the mineral kingdom is
more important to man than iron, and as there is
no metal more widely diffused, it might be supposed
that there could be no difficulty in obtaining it when
and where it is required.   This, however, is by no
means the case, for being itself, as a metal, exceedingly refractory, fusing only at a very high temperature, and mixing in the furnace with remarkable
facility with substances injurious to it, iron is difficult
and troublesome to bring into the metallic form in
a marketable condition, and either requires much
fuel and strong fluxes to separate it from the combinations in which it exists in the more earthy
ores, or else when in its purest state, and combined

only with oxygen, it is exceedingly apt to burn out the walls of the furnace in which it is melted, owing to the fierce heat evolved. While, therefore, some of the purest and finest qualities of iron are obtained now, as they have been from time immemorial, from those crystalline ores which contain iron combined with the smallest quantity of oxygen (such as the Swedish and Russian ores, the Indian, part of the Elban, and some others) by far the largest quantity of the iron used for miscellaneous purposes is reduced from ores greatly mixed with foreign substances, of which clay and lime are the most common.

The ores of iron are, practically, almost limited to the oxides (combinations of iron with oxygen gas) and the carbonates of the oxide (combinations of carbonic acid with the oxide), the former being generally crystalline and pure, or combined only with water, and the latter almost always earthy and impure. From the former are obtained the qualities of iron from which the finest steel is made; from the latter all other kinds of merchantable iron. The former are of three kinds—first, those which consist of the smallest combination of oxygen (*magnetic oxides*), containing when pure nearly seventy-two per cent. of iron, and yielding the finest qualities of steel; secondly, those which contain a little more oxygen (*red hæmatite, specular ore, micaceous ore, and iron glance*), yielding rather more than sixty-nine and a quarter per cent. of

iron when pure; and third, the oxides combined
with water (*brown hæmatite*), containing only fifty-
six per cent. of iron.   From the first group of ores
iron is made by a simple process in small furnaces
but at considerable cost, and the other two kinds
are chiefly valuable for mixing with the poorer and
earthy ores in furnaces of large dimensions, where
all the commoner kinds of metal are produced.

The carbonates, like the oxides, present many
varieties, but they are much more generally impure.
The crystalline or sparry carbonates, if without
foreign substances, would yield forty-five per cent.
of iron, but they are almost always mixed with carbonates of lime and magnesia.   It is much more
common to meet with them in the earthy state as
clay ironstones, in which form, indeed, they are
really the most important group, being the most
abundant and the nearest to the coal, of all the
British ores.

The known resources of England for iron ore, in
the year 1851, were thus described by Mr. S. Blackwell of Dudley, who collected with great care a
series of specimens for the Great Exhibition of that
year in London.   " The carboniferous, or mountain
limestones, of Lancashire, Cumberland, Durham,
the Forest of Dean, Derbyshire, Somersetshire, and
South Wales, all furnish important beds and veins
of hæmatite; those of Ulverston, Whitehaven, and
the Forest of Dean are the most extensively worked,
and seem to be almost exhaustless.   The brown

hæmatites and white carbonates of Alston Moor and Weardale also exist in such large masses that they must ultimately become of great importance." The coal measures, however, were then the chief sources of supply, and "the ironstone beds in them, especially in South Wales, Staffordshire, and neighbouring counties furnish the greater part of the iron produced in Great Britain." "The new red sandstone furnishes in its lowest division beds of hæmatitic conglomerate. In the lias and oolites are important beds of argillaceous ironstones now becoming extensively worked, and the iron ores of the greensand of Sussex, once the seat of a considerable manufacture of iron, will in all probability again soon become available by means of the facilities of railway communication." In 1850, the gross annual production of iron was estimated to have been upwards of 2,250,000 tons, divided as follows :—South Wales 700,000, South Staffordshire and Worcestershire 600,000, Scotland 600,000, various smaller districts 350,000. The increase of production has been as follows : in 1750, it was 30,000 tons, in 1800, 180,000 tons, in 1825, 600,000 tons, from which time to 1850, the production was nearly quadrupled.*

In the year 1850, the clay ironstones of the lias and oolites were hardly beginning to enter into important use, and they are described by Mr. Black-

---

* Upwards of four tons of average ore would be required to manufacture each ton of pig iron.

well as " of no commercial value, but curious as showing the almost universal dissemination of this important ore." It is chiefly with a view to show the progress of development of these ores that the present chapter is written, but before doing so a more detailed notice of the ores themselves will be useful.

They are of three kinds—two of them consisting chiefly of earthy carbonates, and the third of hydrated peroxides, the former differing from each other in being associated in the one case with clays (silicates of alumina), and in the other with lime-stones (carbonates of lime).

The earthy matters are mechanically mixed as impurities with carbonates of the oxide of iron, and are sometimes partially, sometimes nearly alto-gether absent, the ore of iron being then often mixed with coaly matter, and becoming what is called a *black-band ironstone*. These black bands pass into impure coal, and are almost confined to the coal formations, though some have been found in the oolitic deposits. They are rich, easily smelted (combining fuel with ore), and are, there-fore, very valuable when abundant.

The whole secondary series of rocks, from the new red sandstone to the Oxford clay of British geologists, both inclusive, contain deposits of irony material, generally in bands, all capable of being used with advantage in the furnace, although many of them have no appearance of being ores, and

have long been overlooked.   Those chiefly used
hitherto are from the lias and the lower beds of
oolite immediately above the lias.   They seem to
have been all originally grey earthy carbonates of
iron, with lime and a little silica, some parts having
been subsequently altered by weathering into
hydrated peroxides.   They are sometimes nodules
in other beds, but often thick uniform deposits.
They always contain a little phosphate of lime and
magnesia, and sometimes the quantity of phosphoric
acid in the black band carbonates rises to between
one and two per cent., while in the grey carbonates
and altered hæmatites of the oolites it occasionally
has the very large proportion of three per cent.
Although so much phosphorus is considered a disadvantage, there are few of the ores that cannot
be used.

The deposits of this kind are so very abundant
and widely opened, that, although they now supply
an important part of the fifteen or twenty millions
of tons per annum required for iron making, their
quantity is considered as practically inexhaustible.
They form thick beds, covering miles of country,
close to the surface, and they are placed conveniently
in the line of the principal railways, which cross
them in various places, so that they can be carried
at very small cost.

The district in Yorkshire called Cleveland is one
of the most remarkable in reference to this new
and important source of supply for iron ore.   It is

T

a tract of low hills, not three hundred feet above the sea, situated in the north-eastern part of the county, and occupied geologically by the lias and new red sandstone, overlaid by patches of northern drift or gravel. It is chiefly the middle lias that at present yields the iron ores, and they occupy a bed seen on the escarpments or steep faces of the low hills, from which have fallen loose blocks. These blocks, which have an irony appearance, owing to long exposure to the weather, are strewn over the beach on the coast, and first led to the discovery of the deposit about the year 1848. On the coast, and also at a small elevation above the flat land between Redcar and Middlesborough on the Tees, there crops out to the surface a solid stratum fifteen feet thick, which at first sight would be regarded as an ordinary sandstone, its external surface somewhat rusted by the oxidation of iron. This deposit, sometimes massive, at others interlaminated with shaly bands, is of a greenish or grey colour, divided by a system of vertical joints, and its structure is generally oolitic, the grains being concentric, and consisting of silica soluble in dilute caustic potash. The greenish-grey colour of the bed is owing to the presence of silicate of iron, well known to produce the same appearance in the sands of the south-east of England, geologically called "green sands," although often grey, yellow, or even deep red. The whole deposit belongs to what is called the marlstone formation, widely spread through England, and belonging to the middle part of the lias. It abounds with fossils, especially *Belemnites*, and a species of *Pecten* (*P. æquivalvis*). The latter is often found in a very fine state of preservation.

This remarkable seam contains, on an average, about thirty per cent. of iron, and extends, with gradually diminishing thickness, southwards to Gainsborough, and then to Thirsk, where it is too thin to be observable. It covers a region of several hundred square miles, being capped by sandy shales also containing ironstone nodules, and ultimately by the upper lias-shale, so well known and extensively worked at Whitby for the manufacture of alum and the extraction of jet.

Further to the west and south, beyond Thirsk, the lias is covered by the lower oolite series, which also contains valuable beds of workable ironstone.

The great lias deposit of iron ore above described has been chiefly worked for the last six years at Eston, near Middlesborough, the workings being carried on to the full height of the seam by removing part of the ore at a time, and leaving massive pillars.

In a careful analysis of the ore made in the Government School of Mines by Mr. A. Dick, the yield of a specimen, after being dried at the boiling point of water, was as much as 33·62 per cent. of iron. The analysis is as follows:—Protoxide of iron, 39·92 ; peroxide of iron, 3·60 ; alumina, 7·96 ;

lime, 7·44; magnesia, 3·82; silica, 8·62; carbonic
acid, 22·85; phosphoric acid, 1·86; water, 2·97.
There was also a little protoxide of manganese,
potash, bisulphuret of iron, and titanic acid.
Practically this may be regarded as an ore yielding
thirty per cent. of metallic iron.

It will not be out of place here to calculate what
a deposit of this kind may be assumed to yield.
Estimating the average thickness at twelve feet,
and supposing that eighty per cent. can be obtained,
we shall find that each acre of ore-ground will
yield about 40,000 tons of ore,* from which 10,000
tons of metallic iron may be manufactured. Every
square mile, therefore, of such ore may be estimated to yield more than 6,000,000 tons of iron,
equivalent to a year and a half's consumption for all
England at the present rate of manufacture. The
several hundred square miles of iron ore existing,
or supposed to exist, in the lias of the Cleveland
district only, would thus of themselves be sufficient
to feed our furnaces in that part of England for a
period amply long enough to justify us in the free
use of iron.

But the supplies of iron ore from the beds of the
secondary period are, as we have seen, by no means
confined to the lias of Yorkshire. Similar deposits
occur in Lincolnshire, Rutland, and Northampton-

---

* At Eston the beds actually average 50,000 tons to the acre,
and near Whitby, where greatly reduced, as much as 20,000
tons.

shire, where they range also into the oolites, and where they have already been examined and are found workable.   In all the general condition of the ores is the same, and, in addition to those averaging upwards of thirty per cent., there are others far more extensive whose yield would not at present justify expensive operations, but which hereafter, if iron should become dear, or the cost of transport be sensibly reduced, would immediately rise into importance.   Besides, there are rich and very extensive deposits in hollows of carboniferous and permian rocks already largely worked in Cumberland and Lancashire.   These are red hæmatites.

It must not be supposed that England alone is thus rich in iron ores.   In France and some parts of Rhenish Prussia, where generally the ironstones of the coal measures are absent, the oolites are quite as well provided, and the ores quite as cheaply obtained as in England.   Labour being for the most part cheaper than with us, the ore, even when it requires washing to separate the richer portion from the earth with which it is mixed, is delivered at a very small cost; but the expense of carrying it to the coal is such as to limit very seriously the value of the deposits, and renders the manufactured metal dear.   The ores of iron in France alone are said to be found in a workable state over sixty departments, but the total quantity raised in 1854 was under four millions of tons; and even this was reduced by washing to about half its weight.

" These ores are various in character. They are found stratified in beds and in accumulated masses in the form of grains, of kidney-shaped concretions, and of geodes in cavities of Jurassic limestones, or lying upon them superficially. They are met with in the state of grey earthy carbonates, and in that of hydrated peroxides. Many of the deposits are merely superficial, and only covered by diluvial clay and sands. The hydrated ores resulting probably from the destruction and re-collection of the Jurassic beds extend upwards into the tertiary formations. They are found lying upon the latter in many places in the three northern departments of the Nord, the Pas de Calais, and the Ardennes."*

In Silesia, the supply of iron ore is chiefly from crystalline and earthy ores found in the middle part of the new red sandstone series; and elsewhere in Europe large deposits of iron are also present in those as well as newer beds of the secondary series. Most of these deposits are of a nature which renders it difficult to recognise them by the eye as valuable iron ores, and thus it is only recently that their great value has been at all recognised; whilst probably in many places and among many deposits where its presence in a valuable form is not yet thought of, there may be equally large and rich accumulations that will some day be discovered.

---

* *Journal of Society of Arts* for 1855-6, vol. iv. p. 69. Mr. Blackwell's paper.

This is, however, one of the cases in which a combination of different kinds of mineral wealth is necessary, in order that the riches should be available. Iron ores in themselves are absolutely worthless, and often mischievous, as impoverishing the land for agricultural purposes, unless the means of reducing them to the metallic state are at hand; and thus the supply of fuel, its abundance, pure quality, and cheap extraction, are the best measures of the value of such ores. Thus it is that, though other countries than England may contain supplies of iron ore little, if at all, inferior in quantity, and far superior in quality, not one has yet been found which combines so frequently, and in places so conveniently situated for transport, on a very extended seaboard, the various requisites for profitable working. These may be briefly stated as including sufficient and sufficiently good ore with the requisite supply of good fuel, all obtainable at moderate prices, and associated either in the same mine, or within a few miles of each other, and not far from the additional minerals required for the purpose of reduction. No country is known at present, and none is likely very soon to be found, that can produce metallic iron and manufacture it with the facility and cheapness that Great Britain (England, Wales, and Scotland) is enabled to do. Others may bring into the market of the world particular kinds of iron or iron articles of superior quality and excellence to those produced within the British kingdom; but, on the other hand,

England can compete with every country of the world, not only in the open market, but even against the most outrageous protecting duties short of absolute prohibition; and we consequently find the iron and iron manufactures of England make their way everywhere to an extent whose limit seems at present altogether unattained.

XV.

COAL-FIELDS AND COAL EXTRACTION.*

Quantity of coal raised in Great Britain—Space occupied by it in the earth—Limit of the supply—Value of coal—Distribution of coal lands in England and Scotland—Total quantity present in the earth—Estimated quantity that may be obtained —Foreign European coal-fields—American coal-fields—Prospects of England with respect to exhaustion.

WHEN we are told by competent authorities that some eighty millions of tons weight of coal are every year raised and used within the compass of our narrow island, it is impossible not to feel something approaching alarm as we contemplate the possibility of at least a partial exhaustion of the supply for which the demand is so vast. It is not at all easy to realize the meaning of so large a quantity as eighty millions of tons; but we may approach to some idea concerning it if, instead of mere weight, we reduce it to some other dimension. Let us first see how large a building would be required to house a single year's consumption.

* This article first appeared in *All the Year Round* in the number published on the 17th of March, 1860.

Coal weighs, in the compact state in which it is found in the earth, something less than a ton to the cubic yard; so that in order to contain eighty millions of cubic yards of coal unbroken, our building must cover a square mile of ground, and have a clear height of about eighty feet. In the state, however, in which coal is sent to market, much more space would be needed. In order to bring this coal to our store, which, for convenience' sake, we may consider to be placed in the neighbourhood of London, let us see how the three main lines of railroad coming through coal districts would manage to carry the load. Regarding a train drawn by one engine as carrying about one hundred and fifty tons, and assuming that six such trains could be dispatched every hour, day and night, without intermission, it appears that about a thousand tons could be delivered per hour by each line, making a grand total of seventy-two thousand tons per day. At this rate, however, we should only have delivered, at the end of the year, a little more than twenty-six millions of tons. In other words, not one-third part of the year's consumption of coal could be conveyed to a central point, if the whole business of three complete railways was devoted to that purpose. Or, if we suppose the coal transported by ships and carried by screw colliers, each of a thousand tons burden, and performing the round trip in ten days, it would require a fleet of upwards of two thousand such ships (not allowing anything for repairs and accidents) to carry the coal from the mine to the store.

Next, let us consider how much space this quantity of coal occupies in the earth before extraction. An average thickness of workable coal in a very profitable coal-field is about six feet; but it must not be supposed that the whole of this can be taken. Even under the most favourable circumstances there is a loss of twenty per cent., and it is seldom that any large extent of coal exists without some of those fractures and troubles which greatly diminish its value. It would require, therefore, at least fifty millions of square yards, equal to about seventeen thousand five hundred acres, or not far short of thirty square miles, of a single bed two yards thick, to supply the annual demand.

These are large figures, and may be considered to justify the alarm of some of our legislators, who would have us at least retain the power of checking any greatly increased demand which may arise among our neighbours on the other side the Channel. This is a case in which a little sound practical knowledge of geology is required : lest, on the one hand, we should permit our country to be deprived of the fountain of all her wealth ; or, on the other, we should prevent the carrying on of a fair trade in a raw material which we possess in greater abundance, and can sell cheaper, than our neighbours.

Looking at the question from the first point of

view, we are bound to remark that our share of this kind of mineral wealth is limited. It is a great patrimony bequeathed to England, Wales, and Scotland, by the races that preceded us in the occupation of the country—an inheritance of personal property, if we may be allowed the expression, consisting of capital that can be spent: not like an entail of landed property that we can only occupy; we are, therefore, responsible morally to those who may come after us for the proper use of it. We have no right to waste or destroy it, nor in any way to interfere with the value of what we do not immediately require.

As a property it is difficult—perhaps impossible —to exaggerate its importance. It is at present strictly and absolutely the source of all mechanical power. With it we can do and obtain anything that requires power—locomotion by land and sea, manufactures and manufacturing implements of all kinds—heat, and light. All our domestic arrangements are dependent on it. Without it we should hardly be able to call ourselves a people. We have no other sources of fuel, and, therefore, no other means of obtaining steam, which, at the present day, is a necessity of our existence. And we have no means of replacing from our large profits in the use of it, one particle of this magnificent capital. We can use, but we cannot create it. How coal was formed is still to some extent a mystery; but that it has taken far longer to elaborate than the human race has done to complete thus far its history on the earth, there can be no doubt. If coal be now forming, man is not assisting, and knows not how to assist, in the operation. Nor is there any great probability that large deposits of undiscovered mineral fuel exist near the surface of the earth in any part of our country. Doubtless there is coal, and perhaps in large quantity, under certain of the rocks that have not yet been sunk through. The general limits, however, even of these unseen stores are pretty well known, and they form a reserve which will not be touched till the cost of extraction of known deposits is much increased, or the expense of opening out a coal-field at considerable depth much reduced.

Keeping these considerations in view, we may proceed to consider the extent of our known resources and the prospect they offer of permanence. For this purpose let us estimate the area of country occupied by those rocks amongst which coal may be expected to be found. The districts of this kind are called coal-fields; and in all of them coal-husbandry has advanced pretty rapidly within the last few years.

These districts are numerous and extensive. The most important are thus described: 1. The Newcastle coal-field, in the counties of Northumberland and Durham. 2. The Lancashire, including Flintshire and North Staffordshire. 3. The Yorkshire (East Riding), including Nottinghamshire and

Derbyshire.   4. The South Staffordshire.   5. The
Somersetshire and Gloucestershire, including the
Forest of Dean.   6. The South Wales.   Besides
these, coal underlies parts of the counties of Cumberland, Westmoreland, the West Riding of Yorkshire, Shropshire and Worcestershire, Warwick-
shire and Leicestershire.   There are in Scotland a
number of detached coal-fields, of which that in
the valley of the Clyde and Lanarkshire is the
chief.   Ireland is not without coal, but the quality
is poor, and the position of most of the fields inconvenient.

The Northumberland and Durham district of
coal-fields is a compact area of half a million of
acres, in which as many as eighteen beds are known
which are thick enough to pay for working; but
they are not all present on the same spot, and the
thickest does not exceed seven feet.   It is calculated,
and with some approach to precision, that the
average thickness of coal over the whole field is
about twelve feet (including all the seams), giving
a total estimated content of about ten thousand
millions of tons.   If only one-fourth of this be
obtainable, there should still be two thousand five
hundred millions of tons, of which, perhaps, five
hundred millions are already taken or rendered
unserviceable: leaving thus in this one field about
two thousand millions of tons, or twenty-five years'
supply for the whole kingdom.

But   the Newcastle coal-field is neither the

largest nor the most productive of our districts, although it is the one that has been longest opened. The Lancashire district is as large, and has a far greater thickness of coal. The Yorkshire is larger, but the coal-beds are not so numerous, though some are thicker. The South Staffordshire is small, but the thickness of the coal is exceedingly great, amounting in some parts to between thirty and forty feet. The Somersetshire contains a large number of beds, and the total thickness of the coal is very great, but the area is only one-half that of the Newcastle: while the South Wales, with a much greater available area, has thicker beds, more of them, and altogether a much larger supply. The Scotch coal-fields occupy together at least three times the space of the Newcastle, and the thickness of available coal in them is more than double; the thick seams being also double.

Bringing these figures together, we shall find that the whole area of known coal-fields in England, Wales, and Scotland exceeds four millions of acres, or six thousand two hundred and fifty square miles; that over this area there is an average thickness of coal which cannot be estimated at less than fifteen feet, or five yards; and that, therefore, the estimated quantity of coal is equivalent to a bed whose surface occupies thirty-one thousand two hundred and fifty square miles one yard thick.

The eighty millions of tons annually consumed

at present would be equivalent to an area of nearly fifty square miles one yard thick; and thus an estimate of six hundred years for the duration of our coal, at the present rates of consumption, would seem to be justified.

But there are certain very important deductions that require to be made. One, indeed, has already been allowed for in our estimate, as the actual extent of country shown on our geological maps as coal-bearing amounts to about twelve thousand square miles, and the calculations of acreage made do not much exceed half that amount. Fifty per cent., therefore, has already been deducted for unproductive portions of the fields where the coal is injured and unobtainable, whether from faulted ground, inconvenient depth, or patches of bad quality.

We must, however, make a further large deduction, if we would fairly approach to a solution of the practical question. From the total acreage of coal lands, a coal surveyor, in estimating the value of a district, would deem it fair, not only to strike off fifty per cent. for the injured and faulted coal, and the deep parts of the beds, but he must make a further allowance for what is left underground to support the roof, and for the loss of upper beds when the lower ones are first extracted. Our thirty-one thousand two hundred and fifty square miles of coal one yard thick will thus dwindle down to twenty thousand.

Still, there remains a supply equivalent to four hundred times that which is now annually extracted; but, as all these calculations are made on the assumption that no coal has been removed, and, as our coalowners have been doing their best, not only in the way of fair extraction, but very unfair destruction, for many years, we fear that at least a century more must be struck off from this period, if we would fairly estimate our resources. The consumption, too, is not fixed at eighty millions; and if we go on manufacturing and exporting coal and iron at an increased rate, it is obvious that the annual extraction must increase also.

What, then, is our security that we shall not really be drained of our coal within a comparatively brief period? A few centuries form but a small part of the history of a nation, and Englishmen will hardly be satisfied to feel that the days of their country's glory are numbered, and that if they look forward only just so many years as have elapsed since Elizabeth reigned and Shakespeare wrote, their great patrimony will be spent and their source of power at an end. To satisfy ourselves on this point, we must compare the resources of other countries in this respect with those of our own.

Belgium, France, Prussia (both on the Rhine and in her eastern provinces), Russia, Spain, and even Portugal and Turkey, all possess coal-fields as well as England. Belgium and Prussia are producing countries in this respect; and though they do not compete with England in the open market, they are enabled, by aid of their coal, to undersell us in some branches of manufacture.   France is opening out her coal-fields; but France, like all the other countries of Europe, whether provided by nature or not, is chiefly a consumer of her neighbour's stock. Belgium and Rhenish Prussia are the only countries out of England that really work coal-mines on a large scale.

But not only is there coal thus reserved in various parts of Europe; Asia contains it, Africa has its share, Australia and the islands of the Eastern Archipelago possess large stores, and North America has resources so large and so conveniently situated, that time only can be needed to bring her openly into competition with England on very favourable terms.   For every square mile of coal-field England contains, North America contains at least twelve; and for the most part the North American coal is thicker, more easily worked, and a larger proportion of the whole would be obtained.

So far, then, as the world is concerned, there is no fear that coal will perish out of the lands. Parodying the words of our great laureate, we may say,

Men may come and men may go,
But coal burns on for ever.

Practically, there is no fear of exhausting the

patrimony which nature has been storing up for man during countless centuries; and we may even greatly increase the general consumption without danger, so far as the interests of mankind are concerned.

But still the question recurs, How is England affected? To this question the reply is brief and satisfactory. So long as England can raise and sell coal, and make iron cheaper than other nations, so long will her coal-fields be the chief sources of supply; and there is no good reason why they should not be. The day, however, will come, and cannot be far distant, when a continued demand will enforce a more costly mode of extraction, and the price of coal—and, as a necessary consequence, that of iron, of all means of transport, and of manufactures—will rise also. Up to a certain point the different people who purchase our coal, iron, and manufactures will pay the increased price; but, as the gradual exhaustion of our resources renders the remainder more expensive to obtain, the time must arrive when our present customers will use their own coal, make their own iron, and, to a certain extent, manufacture for themselves, or buy in a cheaper market. The exhaustion of our coal-fields will thus be indefinitely delayed, as there will be amply sufficient for our own purposes at prices which, though higher than at present, will not make more difference than to stimulate our ingenuity, and induce future discoverers to find some substitute for coal, in regard to many purposes for which that kind of fuel is now largely used.   Even should we find it economical to import coal for certain purposes, there is no need to fear that we cannot employ our people with advantage, and retain that position among the nations which we have succeeded in gaining.   In North America, in India, and in Australia we have children who, while they profit by their own wealth, will, with advantage, interchange productions with us, and so long as the old English feeling prevails, there will be no difficulty in finding the right direction for English industry.

XVI.

GOLD DEPOSITS.

*Discovery of gold in California and Australia—Action of water
in producing gold alluvia—Three kinds of mining operations
carried on in Australia—Comparative poverty of the lower
part of the quartz veins—Reasons why gold-quartz mining is
unusually speculative—Geology of gold mining—Natural history of gold deposits—Calculation of the quantity of gold obtained in comparison with other metals, and the space it occupies—Relative value of gold, iron, and coal, and comparative
estimate of the hands employed.*

In the year 1847, as every one knows, gold was
discovered by accident in one of the smaller valleys
of a river running down from the Rocky Mountains
to the Pacific, and four years later similar discoveries
were made in Australia. From that time to the
present gold seeking, and, it may be added, gold
finding, have been among the common employments
of the Anglo-Saxon race; one important result
having been to introduce and pour into every
corner of the civilized world a continuous stream of
the precious metal, and another to find out in
succession fresh localities whence this stream could
be supplied.

All the gold hitherto obtained in these new localities has been associated more or less with quartz, and the greater part of it has, before being found, become mixed with fragments of stone in the state of alluvium or diluvium; in other words, the gold has been, by some process of nature, broken away from the rock or vein in which it was elaborated, and carried some distance by water. Water, in fact, acting in its usual way, has broken up and reduced to fragments the rocks and minerals it has acted on, and afterwards has arranged them again nearly in the order of their specific gravity. Then, on reaching the lower parts of each mixed deposit belonging to any one epoch, where, of course, the gold, as the heaviest substance, has always sunk, the rich and sparkling prize is obtained by the miner, either in fine particles not larger than grains of sand, or in larger lumps— or *nuggets*, some of which are many pounds in weight.*

In California the gold alluvium has been found at various depths, but always at or near the bottom of a deposit of detritus or local gravel, which covers slates and schists. The gold is associated with clay or sand, but generally there is little found, except in near contact with older rocks, and the richest

---

* It is clear that this same action of water, which occasionally buries the gold in the deepest part of an alluvial deposit, may, if it has acted subsequently, sweep away all the lighter material, or stone and gravel, leaving the gold nearly uncovered. This case is exceptional, but not unknown.

deposits are at the opening of ravines and in the middle of gullies.

The underlying rocks are described as consisting of gneiss, crystalline limestone, mica slate, chlorite slate, coarse felspathic slate and clay slate. All these are usually vertical, and are traversed by quartz veins. The range of the belts of rock and quartz veins seems to be frequently the same; and, at any rate, the principal veins containing gold are those which are most usually parallel with the underlying crystalline rocks.

In Australia, in Victoria colony, the gold has also been found chiefly in the drift, which there attains a great thickness and ranges over every part of the country. The underlying rocks there consist of sandstones and grits, alternating with clay slates and quartz rock, and are referred to the oldest or palæozoic period. This drift is found in the gullies, on the flats, and capping the hills, and the fragments of which it is made up are not much water-worn, and do not seem to have been transported far. In the Victoria colony, however, the gold-bearing drifts are described as belonging to three distinct periods, reaching back into rocks of the middle tertiary age. These latter occupy very wide tracts of country, and are not unfrequently covered up by a more modern deposit, consisting of lava poured out from a volcano. The ancient drifts are, in other parts of the colony towards the coast, replaced by regularly-bedded deposits of the same period, full

of fossils, also covered with the basaltic lava. Over
this again is another series, partly consisting of
more modern drifts yielding gold, and containing
bones of extinct and living quadrupeds, and partly
made up of more regular fossiliferous beds. Not unfrequently these two gold-bearing beds overlie each
other without the intervention of the lava, and occasionally they are covered by a third gold alluvium
still more recent. There are thus, as we have
said, three distinct gold-bearing strata, occasionally
lying one above another in the same locality, and,
as it is a general rule that the heaviest deposits and
largest nuggets are found at the base of whichever
of the three happens at any given point to repose
on the old rocks, so the richest deposit of all is
almost always at the bottom of all.

It is evident, therefore, that in Australia the
deposit of gold in drift in very large quantities is
an operation that has been going on for a long
time, probably through the greater part of the
tertiary period. Whilst these gold alluvia were
being formed, the volcanos in the neighbourhood,
now apparently extinct, were in full activity, and
floods of lava were occasionally poured out over the
partially-accumulated material, burying much of
the rich gold gravel and preserving it from further
mischief. Who can tell how long it may be before
these deposits are obtained, and what revolutions
may take place in Australia before the hidden
treasures of this field are brought to light?

It would seem, however, that the veins or rocks from which the gold was removed to be buried under lava, were not exhausted. Other and newer deposits, also rich in gold, were heaped on the same bed of lava that had covered the first deposit; and, owing to local causes, not now determinable, a large accumulation of gold detritus and gold gravel took place not only after the first lava had been poured out, but even over this second and newer heap, covering the whole, and now to be reached nearer the surface.

It is chiefly in the Bendigo and Ballarat diggings that these curious complications of the gold drift have been observed; and here also the country has been so much subjected to the eroding action of water, that it is cut up into numerous gullies and watercourses, laying bare in different places the different auriferous deposits at various depths.

Thus in this part of Australia there may be for a long time three kinds of mining-pits sunk, one for the purpose of reaching the lower gold gravels beneath the lava, another belonging to the ordinary surface operation of streaming, and the third connected with an occasional search for the quartz veins from which all the gold was originally derived.

There is no reason why such works may not go on for an extremely long period, if, as seems likely, the deposit is of an average richness through all that part over which the lava has flowed, and if also the quartz veins are sufficiently rich to pay for

working.   At present they appear to pay to a
depth of 50 or 60 fathoms, but they are said to
become less rich in descending.

Like all metals found in what is called the native
states—that is, nearly pure, and mixed only with
other metals, not with oxygen, carbon, or sulphur—
gold occurs in the vein very irregularly, here and
there in great quantity, and in the intervals
totally absent.   It has been supposed, and hitherto
experience seems to have confirmed the view, that
the upper extremity of the quartz veins near the
surface has been so much richer in metal than the
lower part even at a small depth, that no operations
of deep mining for the quartz with gold embedded is
likely to be profitable; and this is not improbable,
owing perhaps as much to the total want of anything
like an average value of rock, as from the increasing
poverty of the vein.   All mining is speculative, and
subject to sudden alternations of success and failure;
but generally there is less speculation in the case of
poor ores, of which large quantities occur together,
indicated by known conditions of the ground above,
than when the ores are more valuable and in
smaller quantity.   Thus, in copper, the rich and pure
native copper worked recently on the shores of Lake
Superior, may be said to be the only successful instance in the world of mining operations for the
native metal on a large scale; and even here the
success is far inferior to that of the Burra Burra
mines in Australia, the Cobre mines in Cuba, or the

great Devon Consols Mines in Devonshire, where the ores are comparatively poor, but where also the average produce is well kept up. This is still more the case with native silver; and in gold, up to the present time, the yield from quartz reefs has been so precarious that, notwithstanding the enormous supply from California and Australia, no associated labour involving foreign capital has been found to answer.

The geology of gold mining is generally of the simplest kind. In Siberia the gold alluvial beds contain the bones of the mammoth, the ancient elephant of northern Asia, and are thus traced to belong to the drift period. In California no fossils appear to have been discovered, but the age is probably about the same. In Australia, as we have seen, the date of the gravels traces back much longer, and involves a much more ancient part of the tertiary period. But the veins themselves, whether of quartz or iron pyrites, from which the gold has been removed, are far more ancient. They are found in rocks of the oldest geological period, and if not actually contemporaneous with them, are certainly of no modern growth.

Their origin is not easy to trace. Gold does not combine with sulphur, at least in any known method; and this, which would have seemed the simplest and most natural mode of introducing the metal into the quartz rock of the veins, or into the pyrites, is hardly a justifiable assumption. Once locked up with the rock, whether with iron and sulphur, or alone, one can easily trace it in its subsequent course.   Broken off by the weathering and wearing away either of the vein itself, or the softer rock enclosing it, and those portions first separated which contain the most gold, it will be evident that the specimens having greatest specific gravity would be left behind in the drift, the rest being carried further, in any transit by water, in proportion as they are poor in metal. The rich lumps, also, being once buried, would have no great tendency to remove further, or suffer other change than is produced by the gradual reduction into smaller fragments of the enclosing rock. Gold is one of the least destructible of all substances by exposure ; and thus the quantity, however small in comparison with other minerals, would remain and accumulate nearly in the same place for a very long time. Thus have been produced those irregular heaps known in the gold mining districts by the names, more expressive than elegant, of "leads," "runs," "gulches," "gutters," &c., with which mining language has recently been enriched.

Most of these, we are told, take their origin in the neighbourhood of some quartz "reef" (vein), from which no doubt the drift now yielding alluvial gold has been derived. In particular cases, where "runs" are known at various depths underlying each other, they are not generally in the same direction, showing that different causes produced the drift at the various periods.*

The total quantity of gold obtained from the earth, although having a value of so many millions sterling, is not, after all, very large when reduced to cubic contents. If, as seems to be the case, as much as is worth thirty millions of pounds sterling are sent every year from the various gold districts, this quantity of the metal would not weigh more than three hundred tons, and might, if piled together, cover the floor of a small room measuring twelve feet by ten, about three feet high. From the great Devon Consols Mine alone the average of copper ore raised amounts to about eighty tons per day; and the relative weight of ore of this kind would be such, that one week's yield would fully equal in cubic dimensions all the gold raised in Australia and California together during a whole year. The quantity of iron manufactured in England every day is three thousand tons, obtained with great labour, and with the consumption of fifteen thousand tons of coal, from some ten thousand tons of ore. A strange contrast this to the acquisition of the one solitary ton of gold a-day, by

---

* Mr. Rosales, in a note published in the *Quarterly Geological Journal*, 1858, p. 543, gives a sketch of three such runs, the oldest, that at the bottom, bearing from the south-east to north-west, being crossed 140 feet above by a second nearly at right angles to it; and this, again, still 30 feet higher, by a surface channel from east to west. All these contain gold alluvia.

washing sand and gravel, and pounding quartz rock, in California and Australia.

And if we consider relative value, the contrast will be hardly less striking. The value of the gold we have assumed to average thirty millions of pounds sterling. The value of the coal raised from English mines is not much less than this at the pit mouth, even without any expense of transport. The value of the iron before its manufacture, and in the simplest metallic state, is upwards of ten millions.

When it is considered that coal and iron are only beginning their course in these states, and may be regarded as raw materials, the iron capable of being brought into various processes of manufacture, by which its value is raised many hundredfold, while the mineral fuel goes forth through the length and breadth of the land as the source of England's power and mechanical superiority over the world, we may well be content with the share of mineral wealth which nature has allotted to us, and remain content to import the so-called precious metals, satisfied with our ample supply of those more truly precious minerals that provide means of employment for the millions of our working population.

It may be estimated that the population of full-grown males directly employed in obtaining the gold does not approach half a million. This population, no doubt, involves a large number of others of various employments to provide for their various wants, and has been the means of adding, with almost miraculous rapidity, to the population and wealth of the colonies where the gold is found. Beyond this, and the advantage to the world of having an increased supply of a substance so convenient as gold to represent wealth, there seems but little gained by the discovery.

XVII.

WATER-GLASS AND ARTIFICIAL STONE.

Discovery of water-glass in 1825—Mode in which it was obtained —Its properties— Its solubility in water—First idea of its use— Kuhlmann's re-discovery—Ransome's views and their result —First manufacture of Ransome's silica-stone—Improvements —Strength and durability by the ordinary tests.

FIVE-AND-THIRTY years ago an ingenious German chemist, Dr. J. Fuchs, of Munich, published in a well-known scientific journal (Karsten's *Archiv*), a memoir announcing a new product from silica and potash, which he at the time considered applicable to various practical purposes, but which, after a few experiments, was neglected and almost forgotten. This supposed new product was, in fact, a soluble glass, and for that reason has been since called *water-glass*. It was originally made by mixing and melting, over a strong heat, fifteen parts of pure quartz sand with ten parts of crude potash and one of powdered charcoal.* The resulting mass, when melted, becomes a hard, blistered, greyish-

---

* The charcoal decomposes and expels the sulphuric acid contained in the potash.

black glass, which, after cooling, is pounded, and is then found to be soluble in about five parts of boiling water. When this is dissolved, it may be preserved unaltered in a fluid state, in closely stoppered vessels, or may be evaporated into a gelatinous mass, and so preserved in cases made of tin plate; or, lastly, by treating it with alcohol, which throws down a gelatinous precipitate, rapidly passing into a solid mass, it may be obtained in a form which requires no shelter. Water easily and completely dissolves either preparation. A somewhat, but not exactly similar substance is obtained from fusing quartz sand· with soda salts; and a third, also very little different, is made from a mixture of potash and soda. All these substances differ from the well-known *liquor silicum* of· chemists, both in composition and properties.*

The water-glass of Dr. Fuchs has some curious properties besides that of being soluble in boiling

---

* *Liquor silicum* is a *hydrated monosilicate of potash;* the monosilicate being prepared by fusing silica with an excess of carbonate of potash (K O, 4 Si O^2, composed of 31 parts of silica with 69·2 parts of carbonate of potash), and hydrated by dissolving the fused compound in water. This solution is transparent when pure, and has a strong alkaline taste and reaction. It is corrosive. Exposed to the air, it absorbs carbonic acid, and is converted in the course of a fortnight into a transparent jelly, which gradually contracts, and after some months is hard enough to scratch glass. The natural minerals *opal* and *hyalite* readily yield the un-hydrated monosilicate, as potash enters into their composition, and also into that of common flint. The soluble or water-glass of Fuchs is a *tetrasilicate of potash* (K O, 4 Si O^2) ; but, after being treated with alcohol to preserve it in a solid state, becomes an *octosilicate* (K O, 8 Si O^2).

X

water. When it is exposed for some time to the action of the atmosphere, it undergoes a slow change, attracting carbonic acid, becoming opaque and white, and exceedingly hard. If of any thickness, this film cracks and is unsightly. Besides carbonic acid, other acids decompose the solution and separate silica in a gelatinous form, while even the solid glass is affected by dilute acids, silica being separated in the form of powder. Alkalies produce a sort of coagulation of the mass, by means of a partial decomposition.

One of the remarkable properties attributed to this new water-glass, was its capacity of cementing certain solid bodies, chiefly salts of lime, when mixed with them in a state of powder or grains of various sizes. "Under these circumstances, the mixture becomes a kind of cement or concrete, and binds firmly to wood. Oxide of zinc (zinc white) and magnesia also become solid and hard when similarly mixed. The former mixture (zinc white with the water-glass), if brushed in a thin layer upon objects, adheres firmly to them, and gives a good coating, to which colour may be added. All substances impregnated with the water-glass are, however, subject to subsequent efflorescence of carbonate (sulphate?) of soda, always present in commercial potash as an impurity." Another property of the soluble glass is, that it imparts considerable hardness to porous bodies made to absorb it. That made with soda appears to be more easily absorbed than that made with potash.

The idea of Dr. Fuchs was, that by treating with his water-glass the layers of mortar used as the foundation or ground of fresco paintings (some years ago so largely practised in Munich, and elsewhere in Germany), a great advantage would be gained in the increased durability of the whole picture. The celebrated artist Von Kaulbach, and M. Echter, both thoroughly familiar with the art and practice of fresco-painting, appear to have adopted the method suggested by Dr. Fuchs, and experimented with much success as to its various modes of application. We are not aware whether it is followed in fresco paintings of modern date, although it has been recently stated that one of our own artists is about to adopt the method on a large scale.

In the year 1841, sixteen years after the publication of Dr. Fuchs's discovery, appeared the first of a series of papers by M. Kuhlmann of Lille, which were continued till 1857; and were then published in a pamphlet entitled " Silicatisation, or Application of Soluble Alkaline Silicates to the hardening of Porous Stones." To this pamphlet was appended the report of a commission appointed by the French Government to examine the methods of M. Kuhlmann, and their probable value. We shall refer to these memoirs and the report in

x 2

a subsequent essay, " On the Preservation of Porous
Stones;" but allude to it here to show the small
extent to which Dr. Fuchs had succeeded in making
known his discovery.   M. Kuhlmann appears to
have re-invented almost all the applications of
Dr. Fuchs, adding to them, however, several others.

In England the same subject was taken up, in
1844, by Mr. Frederic Ransome of Ipswich, who
was equally unacquainted, it would seem, with
what had been done by Dr. Fuchs at Munich and
M. Kuhlmann at Lille.   Mr. Ransome's notion*
seems to have been that, as the hardest and most
durable of stones used for constructive purposes
were those containing the largest proportion of
silica (a proposition only true in a partial sense, the
real hardness depending more on the mode of
cementing the grains and the material for this purpose than on the grains themselves), he might by
some means obtain a material free from the defects of
terra cotta and other artificial stones of which clay
was an essential and large ingredient.   His experiments first led him to cement grains of sand with
glass, by exposing the mixed sand and powdered
glass to a furnace till the glass was melted ; but it
soon occurred to him that, by substituting a concen-
trated solution of glass (silicate of soda or potash),

---

* See his memoir on "Soluble Silicates," read at the late
meeting to the chemical section of the British Association, held
at Aberdeen in 1859, and published in the *Journal of the Society
of Arts*, vol. vii. p. 758.

he would more readily and completely obtain his object. He, therefore, not knowing anything of Dr. Fuchs's solution, suspended ordinary flint stones, (such as are found abundantly in the gravel of the east of England and in the chalk pits,) in wire cages, inside a high-pressure steam boiler charged with a strong caustic solution of soda or potash, and subjected the whole to a steam pressure of sixty to eighty pounds per square inch. He found that, under these circumstances, the flints rapidly dissolved, and that a neutral silicate of potash was readily produced; and making use of it as a cement, he tried the effect, when sand and small stones were worked by its means into a kind of paste. The stone thus made was found, when dried, to be excessively hard, close, and uniform in texture, and capable of being moulded into any desired form; but, on exposure to water or a moist atmosphere, it gradually became soft and was easily disintegrated. To remove this practical difficulty, he next subjected the moulded stone to bright red heat in a kiln, and then found that his cementing silicate parted with some of its free alkali; and that this portion, combining with part of the sand, produced an insoluble glass, which he believed to be altogether unaffected by exposure of any kind, and which cemented the particles of stone together. The stone thus produced was indeed porous, and open in its texture, but in that state was admirably adapted for making into filter stones, and rubs or whetstones for scythes, for which purpose it has been largely used; and by very simple means and employing mechanical pressure, a much more compact material was soon made which closely imitated some good qualities of building stone. Encouraged by this, he next made small diamond-shaped slabs, or *quarries* for pavement, garden vases, balusters, and other decorative objects, often constructed of terra cotta; but it was soon found, when thus employed on a large scale, and in exposed situations, that the surface became unsightly, owing to the efflorescence upon it of a salt resembling that often seen on damp decaying walls. The crystals thus exuding were examined, and found to be sulphate of soda, a well-known salt, attracting moisture rapidly from damp air. On examining the stone, it was found perfectly sound; and on further investigation it was clear that the sulphate of soda had existed as an impurity in the soda-ash used in the boilers to dissolve the flint, and was increased by a similar impurity of the lime used in rendering the solution of soda-ash caustic. After some trouble, this unsightly and disfiguring appearance was quite got rid of by treating the caustic solution of soda with caustic baryta before putting it into the boiler. The caustic alkali being rendered pure, there was no further trouble on this head.

A stone being thus obtained which could be moulded into any required form, and afterwards kiln-burnt into a hard, enduring building material with exceedingly small shrinkage, was evidently well adapted for those decorative parts of architecture required to be produced in large number at a small cost. Vases and garden ornaments of various kinds, and a superior kind of tombstones, have been largely constructed, and seem likely to take the place of ordinary terra cotta, which they greatly excel in colour, in accuracy of shape, and in the absence of all tendency to become disfigured by damp and vegetation. The stone as turned out of the kiln is remarkably sharp and uniform; but, if required, it can be still further worked by the chisel, like other freestones. It only remained that the strength should be such as to justify its general use; and, by a series of experiments recently made at Her Majesty's dockyard at Woolwich, it has been proved that the power of resistance it offered to steady transverse strain was actually one-fourth greater than the Darley Hill stone, known as a very good kind of building sandstone, while it was three times as great as the best kinds of limestone (Portland and Aubigny), and eight times as great as Bath and Caen. The resistance to fracture transversely being thus satisfactorily proved, specimens of the stone were tried under heavy pressure, and a two-inch cube was found to sustain a crushing weight of twenty-one tons, whilst similar cubes of Darley Dale stone crushed with sixteen tons and a half, and cubes of limestones with much less weight.

These very favourable results of experiment seem fully borne out by the experience of several years of exposure to English winters.   During this time the frost has had no effect whatever upon the most exposed specimens, and the acid vapours of the atmosphere in towns have also failed to injure them in the smallest degree.   The peculiar composition of the stone, which seems to consist of grains of silex cemented by an exceedingly thin coat of glass, is eminently favourable for resisting the kind of attacks to which most building stones are liable.

It is curious that the manufacture of an artificial stone with a silica basis was not effected by Dr. Fuchs, as he appears to have tried, without success, to combine grains of sand into a solid with his solution.   We find, however, that both sand and burnt clay are especially excluded by him from the substances with which his water-glass makes a useful cement.   In all his manufactures of stone, salts of lime were necessary, and he did not expose any of them to a kiln heat.   M. Kuhlmann seems not to have attempted to produce an artificial stone, or, indeed, to have attempted any other use for his soluble silicate, or water-glass, than to wash the surface of porous substances.   So far, then, the discovery of Mr. Ransome appears not to have been anticipated, and is as original as it is valuable.

XVIII.

PRESERVATION OF POROUS STONES AND CEMENTS FROM ATMOSPHERIC INFLUENCES.

Frequent decay of building material—State of our public buildings —Cause of decay—Effect of damp air, frost, and gases in the air—Effect of paint—Patents taken out to remedy the evil— Frequent selection of organic substances, and failure—Trial of water-glass, and imperfect result—Mr. Ransome's discovery of a simple means of decomposing the water-glass, and depositing an insoluble salt on the stone—Result of the discovery.

THE decay of almost all kinds of material used for building in nearly every city in Europe, but especially in the damp, uncertain climate of England, and most especially in the case of the soft porous limestones and sandstones so commonly selected even for the best public buildings, has long been felt and regretted, but has of late years become a subject too important to be much longer neglected. Within a very short time some of the most important specimens of modern architecture, amongst which the Palace at Westminster occupies the first place, have shown symptoms of injury from weathering so serious as to threaten an early oblitera-

tion of the whole of the characteristic features of
the decoration; and no one can have passed through
our principal cities, and examined our cathedrals
and churches, without being painfully aware of the
unsatisfactory state of almost all the stone work.
In many cases, indeed, the stone of modern buildings,
and of the restored parts of those that are of more
ancient date, seems to have suffered more than that
which has been for centuries in its place, even when
the same kind of material has been selected. It need
not be added that in our domestic architecture, wherever artificial stone, composed of terra cotta or
cement, or natural stone of cheap kind, has been
introduced, the same decay has taken place, or is
only kept back for a time by a constant application
of paint—an expensive as well as most unsightly
contrivance.

The causes of decay in ordinary stone and
cements it is not difficult to discover. There is, first
of all, the absorption of moisture owing to the porous
state of the surface; secondly, the constant expansion
and contraction of the moisture within the substance
of the stone, owing to the very wide range of temperature that takes place during every twenty-four
hours; and, thirdly, the occasional exposure to intense cold, when the expansion that precedes freezing
acts with irresistible force, and apparently with such
rapidity as not to give time for the excess of moisture to be driven out as it came in. When it is
remembered that all common stones were originally

deposited in successive layers in water, and that afterwards on drying and parting with their water they must have altered their dimensions, and contracted, cracking irregularly, the ordinary state of such stones will be understood, and the further destruction that will take place whenever the water has penetrated any slight fissures and crevices, and afterwards freezes, will be seen to be inevitable. Certain stones are more crystalline than others, and these on the whole will be the least absorbent, and contain the fewest crevices; but short of picked slabs of fine marble there is no limestone—and there is not any sandstone except quartz rock too hard to be chiselled—that does not contain enough of these absorbent fissures to be dangerous. In addition, however, to the absorption from fissures, all stones formed mechanically are laminated, and most of them absorb at the edges of the laminæ or strata.

Besides the absorption of pure water from the damp air, or from the rain driven by wind against an exposed face of stone, it must be remembered that the air in large towns is always rendered impure by large quantities of carbonic acid gas— the result of so many men and animals breathing in one narrow space—by acid vapours (sulphurous and nitrous), the result of burning large quantities of fuel, chiefly common coal, and the large employment of gas for illuminating purposes—by ammonia, another result of animal life, and also by the enormous quantity of unconsumed carbon in an extremely minute state of subdivision seen in the volumes of smoke that darken the atmosphere of large towns, and hang over the site of the town in a perpetual cloud. Of these, all except the smoke are soluble in water, and are beyond a doubt absorbed by rain as it passes down in drops through the air. All, therefore, are driven against the stone and other exposed surfaces by the wind, and are conveyed into the interior through the pores or crevices of the stone. The inevitable result is a slow destruction of all those surfaces that contain or consist of carbonate of lime—or, in other words, of limestones and marbles of every kind, of all sandstones cemented by calcareous matter, and of all cements, even those that are hardest and most indestructible. Thus we find included among the substances injured by exposure certain granites and porphyries that contain soda among their component parts. In addition to the mischief produced by the direct chemical and mechanical action of water containing acid and alkaline ingredients, is the unsightly effect of smoke blackening the surface, chiefly of course where peculiar draughts of air drive the rain in certain directions, but also in every other part. In London the whitest limestones, such as Portland, soon become sooty black, and those of the best warm creamy tint are rendered as unsightly as the worst varieties of the commonest and worst kinds.

In the case of cements, where the cracking of

the surface and the total destruction of the substance would soon follow complete exposure, it has generally been found necessary to paint the whole, repeating the coat of paint as often as from the circumstances of exposure symptoms of incipient decay appear. Three years is regarded as the longest period during which a stone or stuccoed surface that has been painted can safely be left without renewal in the climate of London; and if it is desired to preserve a surface in a state reasonably white and good-looking, it must be thoroughly cleaned at least once during the period. Where, however, attention is paid to appearances, this expensive process is repeated annually. It is clear that for large public buildings, where it is impossible to reach the exterior for painting without a very expensive scaffold, such a method could never be followed, nor indeed would it be in any way desirable. The effect of paint is only to put off for a time the evil day, and the moment the process of painting is completed, that of decomposition commences.

Very numerous and varied have been the contrivances suggested for so filling up the pores and crevices of stone and cement as to render the surface unattackable by damp air and acid vapours. Many of them have been patented, and there are not less than seventeen such patents bearing date from 1838 to 1857. Of these, eleven depend on the principle of choking the pores of the stone with oily or resinous substances, mixed with blood, glue,

gum, wax, flour, caseous matter, or other preparations, chiefly of vegetable or animal origin, all of
which, when long exposed in thin coats to the air,
become decomposed, and in that case lose their
only use. They must ultimately, therefore, scale off,
carrying part of the face of the stone with them, and
leaving the newly exposed surface to ordinary decomposition. There is no reason to suppose that
in exposed places in London, or any large town
in England, any oxidisable preparation could last
more than a very few years, although there is
little doubt that under certain circumstances of
shelter or uniform temperature of localities partially
sheltered, as the under side of railway arches, &c.,
some of the preparations patented might answer for
a long while. It must be observed, indeed, that
most of them, owing to the dark colour of the
principal material employed, are altogether unfitted
for general use in decorative architectural work for
public buildings.

It has been mentioned in the last chapter that
in the year 1840, or thereabouts, Professor Kuhlmann applied the water-glass, originally discovered
by Dr. Fuchs, to the hardening of porous stones.
He observed that on placing chalk in contact with,
or plunging it into, a cold solution of silicate of
potash, a change takes place, a portion of the
potash being displaced, while part of the chalk,
combining with the silicic acid set free, is converted
into silico-carbonate of lime, and, by exposing the

chalk alternately to the action of the solution and
to the air, he found that the stone in time becomes
hardened to some depth in the interior, this hardness increasing till the altered chalk will even
scratch glass. The process of hardening seems to
be greatly assisted by heat, and is applicable, to a
certain extent, to plasters, cements, and common
building sandstones. The solution is laid on in a
somewhat dilute state, and is recommended by M.
Kuhlmann to be applied to recent buildings by
means of a syringe; but he states that, before
soaking long-constructed and exposed surfaces,
they should be washed and cleaned with a hard
brush and a scraper, assisted by a solution of
caustic potash. He considers that three applications, on three consecutive days, will generally be
sufficient to produce a permanent hardening of the
surface.

While these things were doing in France, an
almost precisely similar method was patented in
England by Mr. Newton, who, in 1841, undertook
to preserve stone by treating it with a solution of
silicate of soda or potash. To whatever cause it
may be owing—whether to the damper atmosphere
of England being less favourable for the hardening
and silicification that are required to give the stone
a hard surface, or for some other reason—the method
does not seem to have been carried out in practice;
but another patent was taken out in 1852 by a
M. Moreau for the same so-called invention, which

was again repeated in 1855 by two other persons—
the principle in all these cases being identical with
that of Kuhlmann.

Meanwhile Mr. Ransome, while manufacturing
on a gradually-increasing scale his artificial silica-
stone, re-discovered M. Kuhlmann's method, and
soon found the soluble silicate to produce great
hardness on stones in dry places, and to be apparently thoroughly effective as long as it was pro-
tected from moisture. On operating with the
solution, however, practically out of doors, he discovered that a shower of rain, or even a damp
state of the atmosphere, at once removed the first
film of hardened material before it had absorbed
sufficient carbonic acid to precipitate the silica and
enable the process to be completed. It is clear,
from an experiment tried, in accordance with M.
Kuhlmann's instructions, on a part of the river
front of the Houses of Parliament at Westminster,
that a precisely similar result has there taken
place; and, although the buildings operated on in
France, and at first favourably reported on, seem
to have been better preserved, owing, no doubt, to
the greater dryness of the climate, and partly, it
may be, to their having been executed at a favourable season, there is no doubt that they are now
beginning to show symptoms of decay, the simple
coating of silicate having failed to produce a permanent deposit excluding the action of the weather.

Mr. Ransome's next step was to try the effect of

a weak acid solution following the silicate, which, by decomposing the silicate and combining with the potash, might set free the silica in the pores of the stone; but, as he found that the silica thus deposited was in a form which gave no cohesion to the particles of the stone, and was liable to be washed out by the next shower, he was forced to seek for some other contrivance.

It then occurred to our experimenter that, if by a process of double decomposition, brought about by two liquid solutions following each other, he could produce not only on the surface, but within the substance of the stone to which it was applied, an insoluble salt of sufficient hardness and durability, and another soluble salt which could easily be removed, the whole difficulty would be surmounted, and silicate of lime fortunately suggested itself as the best mineral to deposit. It is known that the mineral so called is one that more than almost any other possesses the property of *cementing*—a thin film of it adhering more strongly to a grain of quartz, the surface of a stone, or a similar thin film of the same mineral attached to another, than a mass even of the same mineral of any size. It is in fact the cause of the perfect cohesion of mortar, concrete, and other cements, both common and hydraulic, and its peculiarly indestructible nature may be judged of by examining a specimen of Roman mortar after an interval of nearly twenty centuries. If there is any one mineral substance

in common use that more completely resists weathering than any other, it would seem to be this;
and although no doubt it continues to harden as
time goes on, it is at once deposited in a film of
considerable strength.   Mr. Ransome was therefore
right in concluding that, " if it were possible to
form silicate of lime in the structure of the stone
(independently of any decomposition of the stone
itself), so as completely to envelope the several
atoms of which it is composed, he would succeed in
producing a result which would at once materially increase its hardness, at the same time hermetically
closing all the pores, and be lasting in its effects."*

Chloride of calcium (muriate of lime)—a by-
product in the preparation of various salts used
largely in commerce—soon suggested itself as the
material for the second wash, and, on trial, was
found to succeed.   The stone was first treated with
a solution of silicate of soda sufficiently diluted to
be thoroughly absorbed, and afterwards with a
solution of chloride of calcium.

When operated on in this manner, a chemical
combination was found to commence immediately,
the chlorine, parting from the calcium, attacking the
soda, for which it has a greater affinity, and forming
chloride of sodium, or common salt, which is washed
away completely with water, while the calcium
combines with the silicic acid of the silicate, and

---

* Journal of Society of Arts, vol. vii. p. 760.

forms silicate of lime in a tough state, which
attaches itself firmly round the surface of each
separate grain of the stone with which it comes in
contact, producing an extremely compact deposit,
not porous or absorbent of water, nor acted on
either by carbonic or dilute sulphuric acid.

Mr. Ransome having thus succeeded in providing
a means which would most economically, as well as
efficaciously, render durable the most porous limestones, sandstones, or cements, had the opportunity
of trying a fair experiment in the month of
October, 1856, on two of the buttresses of the
river front of the Houses of Parliament, and
though the effect is somewhat unsightly, owing to
a small excess of the silicate of lime inadvertently
used having produced a white deposit on the surface, the real value of the process is abundantly
seen when the parts thus treated are compared with
the rest of the building.

The four winters that have passed have had no
effect on the most decaying portions of those stones
which were properly treated by this ingenious process, whilst in other adjacent stones not so treated
the decay has advanced with more or less rapidity.
In other experiments tried since on specimens of the
principal building stones, some were left untreated,
and exactly similar portions coated.   On immersing
these all in dilute acid, after careful drying and
weighing, and then re-drying and weighing after
immersion, it was found that the stones not treated

with the silicate were quite disintegrated, while those that had been previously prepared were uninjured, and had lost no weight.*

The process that would preserve a porous building stone from decay would act in a similar way, and to quite as great an extent, on cement and plasters, and could manifestly be used to render bricks waterproof. It admits, also, of colours being applied where necessary, care being taken to admit only of such mineral pigments as are soluble.

It should be understood that by this process the face of the stone is not rendered non-absorbent. Desirable as this might appear, there is no reason to suppose that such a condition is at all necessary, and it is certain that Craigleith stone, one of the most durable of all, is by no means in this state. What is required is, that the surface, whether absorbent or non-absorbent, shall not be liable to injury from the expansion of water, or from the acids and other injurious vapours that water, in being absorbed, would carry along with it.

---

* On examining the stone treated by Mr. Ransome in 1856, a considerable efflorescence will be seen. This, however, consists only of crystals of sulphate of magnesia and sulphate of soda thrown out from the interior of the stone owing to the partial decomposition that had commenced before the silicate of lime was put in. The specimens swept off and examined by Mr. Warrington were not found to contain any particles of the stone itself, nor had the decay (which had already continued for some time when Mr. Ransome experimented) in any way advanced either on the surface or in the interior beneath the places where this efflorescence had taken place.

In considering the results of his invention and discovery in respect to water-glass, and the hardening of limestones by silica, M. Kuhlmann, or rather the French scientific commission on the merits of his processes, very pertinently allude to the interesting geological conclusions that arise out of them. Thus the formation of flints, agates, petrified wood, and other silicious infiltrations, is probably due to a slow decomposition of alkaline silicate by carbonic acid, and felspar, with various alkaline and magnesian silicates, have been reproduced artificially by means which help to explain their probable origin in nature.

THE END.

www.ingramcontent.com/pod-product-compliance
Lightning Source LLC
Chambersburg PA
CBHW021459210326
41599CB00012B/1057